지붕, 보호되는 오픈 스페이스
ROOF, protected Open Space

| | |
|---|---|
| 초판 발행 | 2019년 6월 25일 |
| 엮은이 | 담디 편집부 엮음 |
| 펴낸이 | 서경원 |
| 편집 | 나진연 |
| 펴낸곳 | 도서출판 담디 |
| 등록일 | 2002년 9월 16일 |
| 등록번호 | 제9-00102호 |
| 주소 | 서울시 강북구 삼각산로 79, 2층 |
| 전화 | 02-900-0652 |
| 팩스 | 02-900-0657 |
| 이메일 | damdi_book@naver.com |
| 홈페이지 | www.damdi.co.kr |

| | |
|---|---|
| First Edition Published | June 2019 |
| Compiler | DAMDI Publishing House |
| Publisher | Kyongwon Suh |
| Editor | Jinyoun Na |
| Publishing Office | DAMDI Publishing House |
| Address | 2F, 79, Samgaksan-ro, Gangbuk-gu, Seoul, 01036, Korea |
| Tel | +82-2-900-0652 |
| Fax | +82-2-900-0657 |
| E-mail | damdi_book@naver.com |
| Homepage | www.damdi.co.kr |

지은이와 출판사의 허락 없이 책 내용 및 사진, 드로잉 등의 무단 복제와 전재를 금합니다.

All rights are reserved. No part of this Publication may be reproduced, transmitted or stored in a retrieval system, photocopying, in any form or by any means, without permission in writing from DESIGNERS and DAMDI.

정가 15,000원

© 2019 DAMDI and DESIGNERS
Printed in Korea
ISBN 978-89-6801-092-7(94610)

이 도서의 국립중앙도서관 출판시도서목록(CIP)은 서지정보유통지원시스템 홈페이지(http://seoji.nl.go.kr)와 국가자료공동목록시스템(http://www.nl.go.kr/kolisnet)에서 이용하실 수 있습니다.(CIP제어번호: CIP2019021431)

**DAMDI Q&A SERIES 6**

# ROOF
# protected
# Open Space

지붕, 보호되는 오픈 스페이스

# Contents

**006 What is the Roof?
Interview with Architects**

008 What is the favourite roof you've designed and why?

040 What is the least favourite roof you've designed and why?

052 What is the most typical colour of roofs in your country? Is there a reason for that?

072 What was the most memorable roof design you've seen and why?

094 Do you have a certain approach or your own unique touch when designing a roof?
If so, what is the reason? If not, what kind of signature design would you make?

110 Are there any myths, stories or superstitions about roofs in your country?

122 **Architects**    130 **Case Study**

b4 architects
BOARD
ELA(Edu Lopez Architects)
Ezequiel Farca + Cristina Grappin
Katsuhiro Miyamoto & Associates
LGSMA_
Miha Volgemut architect
Moussafir Architectes
NL Architects
object-e architecture
OPARCH
SLOT STUDIO
Stefano Corbo Studio
TOUCH Architect
TROPICAL SPACE

008  What is the favourite roof you've designed and why?

040  What is the least favourite roof you've designed and why?

052  What is the most typical colour of roofs in your country? Is there a reason for that?

072  What was the most memorable roof design you've seen and why?

094  Do you have a certain approach or your own unique touch when designing a roof?
If so, what is the reason? If not, what kind of signature design would you make?

110  Are there any myths, stories or superstitions about roofs in your country?

# Q1
### What is the favourite Roof you've designed and why?

## b4 architect

We designed for a competition a special pavilion for the citizens of Novosibirsk, Russia (The new Summer A&D Pavilion). It was been thought as a prototypic architecture adaptable to different settling conditions dedicated to events to involve people and administration to best understand all the urban processes on the city.

러시아의 노보시비르스크 시민만을 위한 파빌리온 디자인(신규 하계 A&D 파빌리온) 공모전에 참가한 적이 있다. 이 파빌리온은 도시에서 일어나는 모든 변화를 받아들인다는 의미로 관람객과 관계자들에 의해 행사마다 요구되는 다양한 조건에 부응하는 건축의 본보기를 선보이고자 디자인되었다.

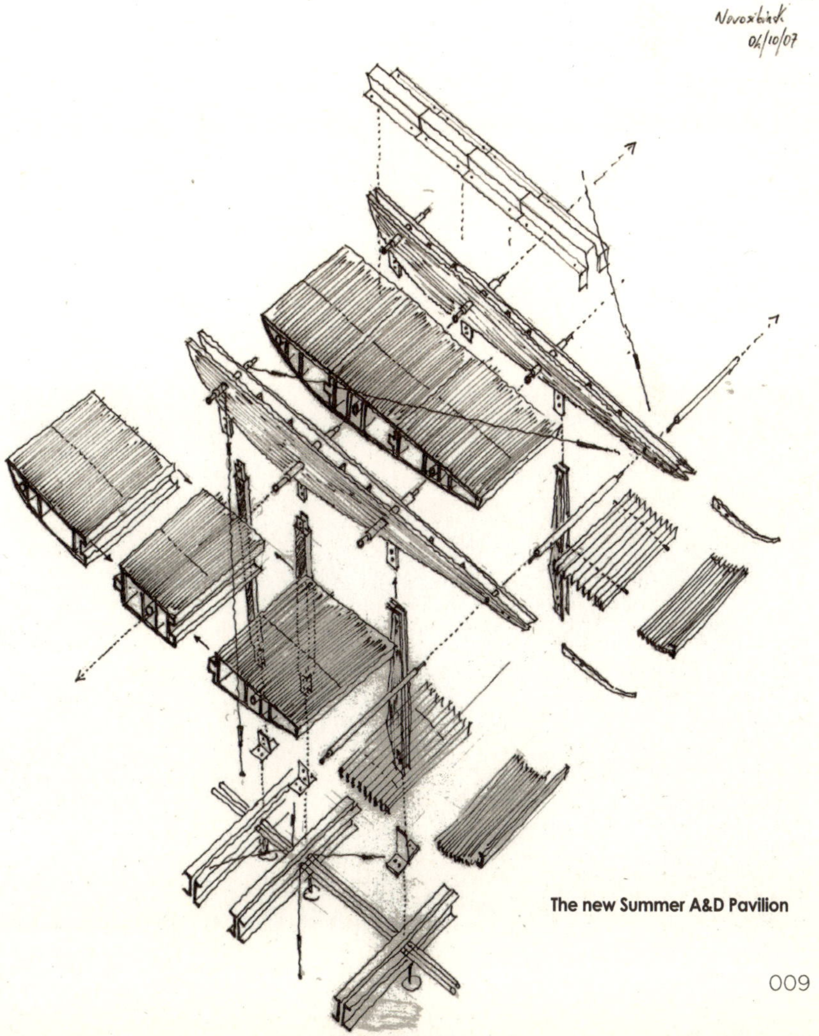

Novosibirsk
04/10/07

The new Summer A&D Pavilion

The strong intent of the project is a mediation between the nature's infinity and the basic human will of building 'inside' spaces. This intent was reached by evoking marine metaphors suggesting the 'floating' building in several contexts proposing itself as clear urban landmark for the community. The role of the roof in this sense was primary: the section and in general the volumetric shape of it remembers a keel of a ship under which all the functions had to be organized. We had the occasion to examine in depth also the construction technology and we studied a prefab system for all the parts, roof included. The whole prefab system was thought to allow the assembly all the components of the pavilion in three days with 4 workers.

이 프로젝트의 가장 큰 목적은 무한한 자연과 "(머무를)내부"공간을 만들고자 하는 인간의 기본적인 욕구사이를 중재하는 것이었다. 이러한 의도로 우리는 지역사회를 위한 명백한 도시 랜드마크로써 다양한 문맥안에서 스스로를 드러내며 해상에 비유되는 '떠 있는' 건축물을 디자인하기에 이르렀다. 여기서 지붕의 역할은 중요하다: 단면과 전반적인 매스의 형태는 배 밑부분의 용골을 연상시키며 이 안에는 모든 기능이 체계화 되어 있다. 우리는 건설 기술을 깊게 조사하여 지붕을 포함한 모든 곳에 적용할 조립 시스템을 연구하였다. 조립 시스템은 4명의 작업부가 3일안에 파빌리온 전체를 조립할 수 있게 끔 설계되었다.

The new Summer A&D Pavilion

# BOARD

For the Dutch city of Hulst we once designed a temporary roof for an open-air theatre that could be easily assembled, disassembled, and stored, together with a temporary stage, temporary tribunes, and 500 chairs in a 3 metre high space of about 40m2 surface area.

네덜란드 도시 헐스트에서 야외 극장을 위한 임시 지붕을 설계한 적이 있다. 이 지붕은 쉽게 조립하고 분해하여 임시 무대, 임시 연단 그리고 500개의 의자와 함께 3m 높이, 40m² 규모의 공간에 보관할 수 있었다.

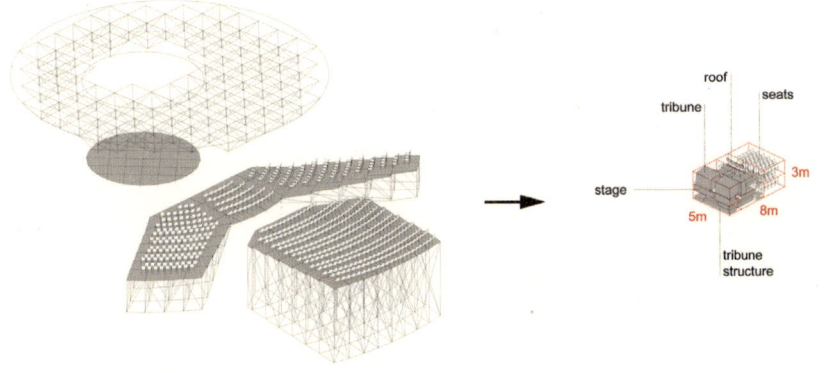

Temporary Elements Stored          Temporary Elements Built

Open-air theatre

011

This made the storage of the roof and the other temporary elements very easy during the winter months when the theatre was closed, and eliminated the need to build an entirely new warehouse. I was pleased that we achieved that by creating a simple steel structure made of thin steel struts. Moreover, the roof can easily be closed in case of rain, with its system of waterproof fabric, and a curtain that surrounds the roof can create a closed interior space around the stage and a small temporary stand for 100 seats.

이런 시스템 덕에 극장이 휴업하는 겨울시즌동안 지붕과 다른 임시 설치물들의 보관이 매우 용이하게 되었으며, 완전히 새로운 창고를 지을 필요도 없었다. 나는 우리가 간단한 철골 구조만으로 해결하였다는 점이 매우 만족스럽다. 게다가 우천시에는 간단하게 방수 패브릭으로 뚫린 지붕을 덮을 수 있으며, 지붕 둘레로 커튼을 둘러서 무대와 100석의 임시 스탠드를 주변으로 닫힌 내부 공간을 만들 수 있다.

## ELA(Edu Lopez Architects)

The project for the expansion of the Prague Congress Center is one of the most complex projects I have undertaken because of its roof. This project called for a multipurpose space of 5,000 m2 and what was decided to do is to activate the greatest technical and formal effort on the deck.

프라하 국회의사당 증축 설계는 우리가 설계한 가장 복잡한 프로젝트 중 하나이다. 그 이유는 지붕 때문이었다. 이 프로젝트는 5000 평방 미터 규모 안에 다양한 시설을 갖춰야하는 공간을 설계하는 것이었고 우리는 데크(지붕) 디자인에 최상의 기술과 정성을 쏟기로 하였다.

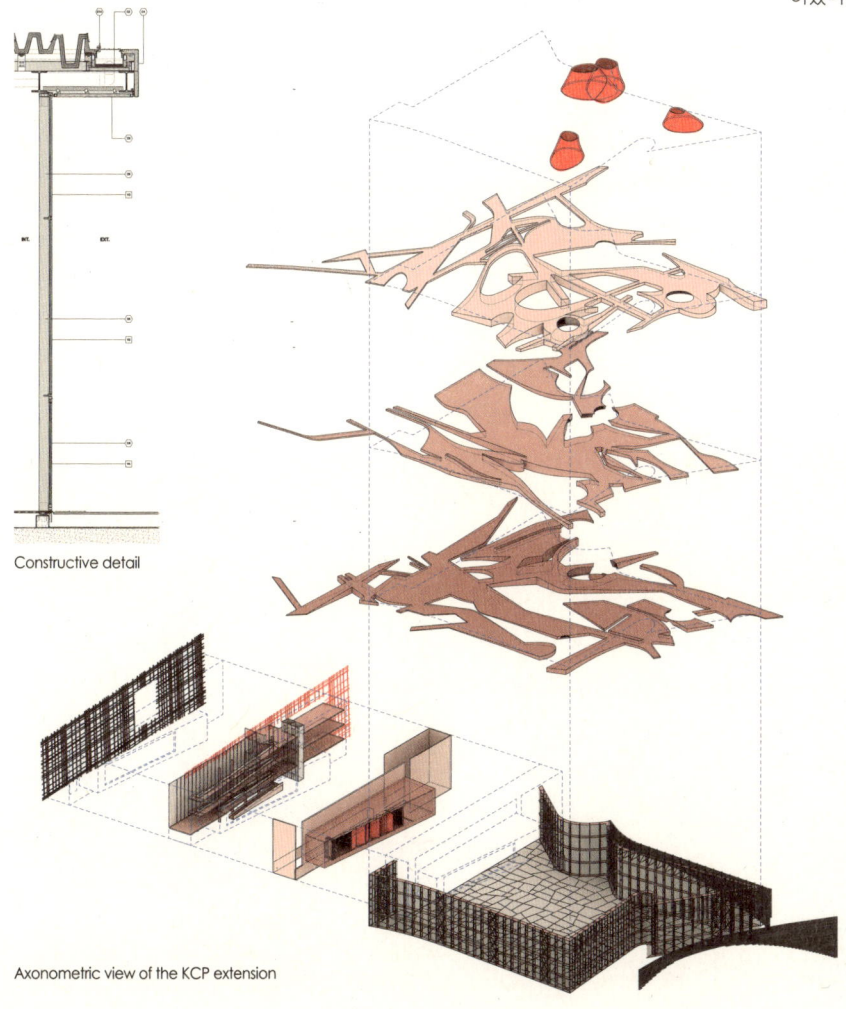

Constructive detail

Axonometric view of the KCP extension

013

The deck becomes the generator of the project, where everything else is based on creating a simple perimeter glass enclosure. Thus, the roof is the protagonist of two horizontal facades, one interior and one exterior. This action of creating this cover so complex geometrically, responds to the needs to visualize an organic element from the inside and the outside, since the height of the new enlargement will be smaller than the main building and the cover will be able to be seen from the interior, creating a formal continuity towards urban planning for the area. It is an element that despite its "controlled chaos, is in unison along with the proposed landscape intervention. The covering an area of approximately 5,000 m2 is based on the superposition of three differentiated layers. Every one of these layers has a geometry in plan and a double curvature in section. Overlapping these three layers creates a multitude of intersections inside the building that make the deck create a certain movement.

다른 모든 요소들을 유리로 간단하게 덮어버리는 건물 디자인에 데크(지붕)는 공간의 활력소가 되었다. 이는 지붕이 건물 외부와 내부의 두 입면에서 주인공이라는 뜻이다. 5000 평방미터 규모의 공간은 세 개의 차별화된 레이어의 중첩에 기초하여 디자인 되었다. 이 레이어 모두 평면상 기하학의 형태를 취하고 있으며 단면상으로는 이중 곡선의 형태를 취한다. 중첩된 레이어는 건물 내부에서 복잡하게 교차되어 다양한 형태의 데크(지붕)를 형성한다.

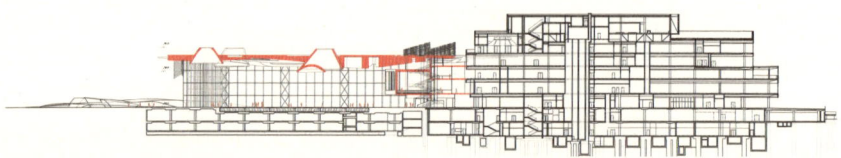

Section view of the KCP extension

In all our projects we tend to flee from the roofs with horizontal flatness and this is one of them that proves it. I do not like that the ceilings do not speak of the formalism of the project, I like that said plane speak a language by itself or allied to the project. Once the three layers of the roof are superimposed, the main structure of the project is incorporated into it, acquiring a maximum thickness of 4 meters and a minimum of 1.5 meters to accommodate the large three-dimensional structure. Being a large space and have a height of 14 meters free. There are large skylights in the form of truncated cones which will give a color tone to the project and shed light into the interior of the space.

우리의 모든 설계작업에서 평평한 수평 지붕은 기피 대상이며 이 지붕이 그걸 증명해준다. 프로젝트의 형식을 반영하지 않는 천장도 좋아하지 않는다. 비행기마다 자기만의 루트가 있다 라는 말을 좋아한다. 프로젝트도 마찬가지이다. 겹쳐진 세 개의 레이어는 메인 구조의 일부로 스며들어 최대 4미터, 최소 1.5미터 두께의 커다란 입체적 구조가 된다. 잘려진 원뿔 모양의 큰 채광창은 14미터의 여유로운 높이의 넓은 실내 공간에 빛을 끌어들여 여러가지 색조를 연출한다.

### Ezequiel Farca + Cristina Grappin

Casa Barrancas is a very special project of home renovation. The location of this, on a ravine allowed us to generate and give the necessary importance to two large terraces. It is one of our favorite projects for the unconventional and special experience that can hold the house, because you find open garden areas on the two upper floors of it, having the possibility of creating totally different

Casa Barrancas는 매우 특별한 주택 리노베이션 작업이었다. 이 주택은 산골짜기에 위치하여, 두개의 큰 테라스의 필요성이 중요하게 부각된 사례였다. 형식에 얽매이지 않은 특별한 경험을 할 수 있었기에 이 프로젝트는 우리가 가장 좋아하는 프로젝트 중 하나이다. 두 곳의 최상층에서 누릴 수 있는 넓은 정원은 완전히 다른 환경 조성이 가능하다. 또한,

environments. In addition to the fact of a low housing, with the benefits of solar radiation and indoor climate that implies.

태양열 에너지로 실내환경을 조절하는 사실상 에너지 절약 주택이기도 하다.

© Roland Halbe

1.- Pool
2.- Principal Bedroom
3.- Social area
4.- Familiar area

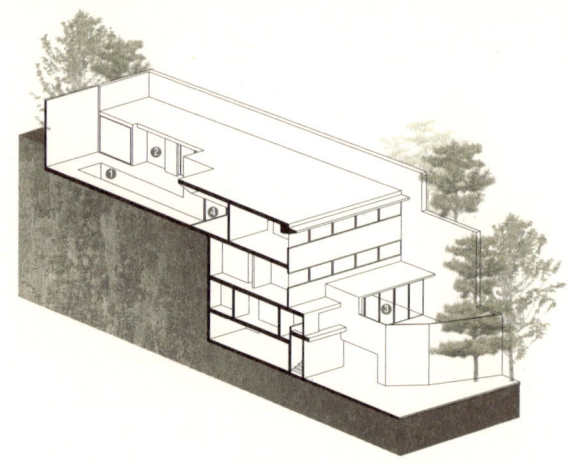

Casa Barrancas

## Katsuhiro Miyamoto & Associates

Chushin-ji Temple Priest's Quarters project has a roof design that was a natural result of function. The form is a response to heavy snowfall, circulation route, seismic resistance and typological tradition.

추신지 절(Chushin-ji Temple)의 승자 숙소 프로젝트의 지붕 디자인은 기능에 순응하는 디자인이다. 많은 강설량, 순환 경로, 지진 내성, 형식적 관습이 반영된 형태이다.

## LGSMA_

The project is composed by 2 buildings. Structure defines architecture and its language in its purest form. The intention is to employ a material that is widely used in construction in Italy, but trying to turn it into a form of expression, by way of a simple construction system and the use of colour. The achieve is a crude, almost brutal, effect that emphasizes its material quality. Concrete is coloured with red pigment - a color which is very typical for the Roman countryside buildings - in the mixture to get an homogeneous material to mould.

The three pairs of rhomboidal section inclined vertical columns are the building's distinctive features. But I hate the idea of the pitched roof. Starting from the same point on the ground two by two, they open up and fold becoming beams of the roof slab; in-between

프로젝트는 2 개의 건물로 구성된다. 구조적으로는 가장 간결한 형태로 건축 성향을 나타낸다. 이탈리아 시공계에서 흔히 사용하는 재료를 사용하고자 하였다. 하지만, 간단한 시공 시스템과 색상으로 형태적인 변화를 주려고 노력하였다. 투박하고 거친 느낌의 결과물은 재료의 질을 오히려 강조하는 효과를 낳았다. 콘크리트는 거푸집 재료와 동일한 느낌을 주기 위해 붉은 안료로 물들였다. 로마 시골지역 건물에서 볼 수 있는 매우 전형적인 색상이다. 마름모 꼴로 기울어진 3쌍의 수직 기둥은 건물의 독특한 부분이다. 하지만 난 개인적으로 기울어진 지붕을 좋아하지 않는다. 기둥은 쌍쌍이 지상 위 동일한 지점에서 출발하여 위로 갈수록 벌어지다가 지붕 슬라브 빔에서 멈춘다. 각 쌍의

the pairs of beams, steel cables allow vegetation to grow, creating a protected porch between the main building and the square. These beams outline a geometric grid, characterizing both roof horizontal surfaces, by way of their red color but also through a small intrados relief (6 cm). The roof slab is supported punctually at each beam intersection vertex by steel columns.

Outer public and semi-private spaces adjoining the building, as well as its interior, are conceived as an extension of domestic space, hosting the spontaneous social initiatives of the inhabitants. As in the local practice of use of land, countryside and urbanity merge together, for an ecologic conversion of territory into a living landscape.

The roof of the second building has been thought as an extention of the public space; it is a square.

빔 사이에 설치된 강철 케이블들은 본관과 광장 사이에 중문 형태의 입구를 형성하며, 케이블을 따라 식물들이 자란다. 이 빔들의 윤곽은 그리드 형태를 띠며 붉은 색상과 작은 내륜 릴리프(6cm) 때문에 양쪽 지붕의 수평면이 부각된다. 지붕 슬라브는 모든 빔 교차점에서 강철 기둥으로 일정하게 지탱된다.

건물 그리고 건물 내부와 접해 있는 외부 공공 공간과 준 사적 공간은 건물 내부가 확장된 것처럼 보이며 이웃 주민들의 자발적인 교류가 이뤄지는 곳이다. 토지 활용 지역 관행에 의하면 토지를 생태적인 생활환경으로 전환하기 위해 시골과 도시가 하나로 병합되는 추세이다.

별관의 지붕은 광장처럼 공공 공간의 연장으로 쓰인다. Ian+에게 모든 건물의 구성요소는 디자인의 기회이다.

© ian+
Hall of Ospedale del mare

Each component of the building for Ian+ is an occasion of design.

© ian+
Hall of Ospedale del mare

### Miha Volgemut architect

Once I proposed to a client a quite daring roof design for his parking lot in front of his house and he accepted the proposal. It is dynamic design added to minimalist house, located at the entrance façade. So, the roof makes the welcoming gesture and gives the character to the entrance of

한 건축주에게 그의 주택 전면 주차장에 대담한 디자인의 지붕을 제안하여 승인된 적이 있다. 미니멀리스트한 주택 전면에 역동적인 디자인을 부가한 샘이며, 반기는 분위기를 자아내는 지붕 덕분에 입구가 개성 있어졌다. 여러 복합물들이 한

the house. It is a satisfaction when more complex structures are manifested in physical space.

물리적 공간안에서 모두 표현될 때 매우 만족스럽다.

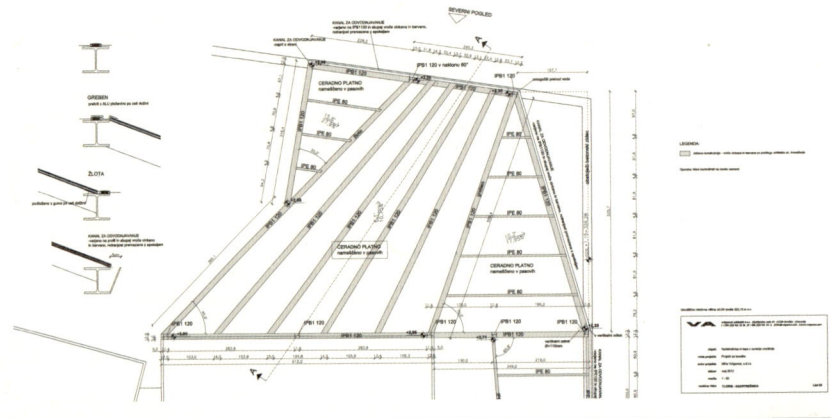

022 ROOF, protected Open Space

## Moussafir Architectes

My favourite roof is the one we designed for the "House in an orchard" located in Montreuil near Paris. The plot reminded me of the Garden of Eden from my first visit. It was running along an old peach-growing wall, forming a green enclosure and delimiting an area of calm, cut off from its urban surroundings. I found it was a place lending itself naturally to human habitation rather than a receptacle for a building.

나는 우리가 설계한 파리 근처 몽트레유의 "House in an orchard" 지붕을 가장 좋아한다. 처음 대지를 방문했을 때 에덴동산이 떠올랐다. 복숭아가 자라는 오래된 벽을 따라 형성되는 녹색 울타리는 도시 환경과 단절된 고요한 영역을 구획한다. 나는 이 장소가 건물을 위한 대지이기 보다는 선천적으로 대지 스스로가 인간의 주거지가 되는 곳이라고 생각하였다.

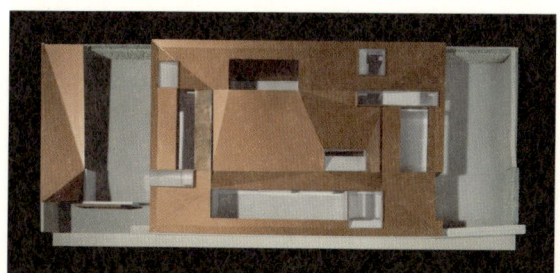
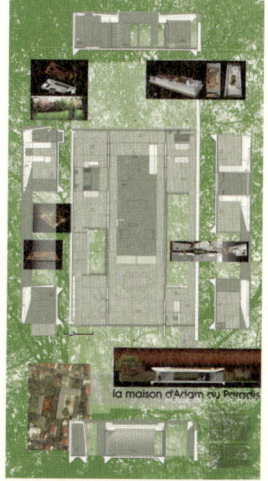

© Gilles Coulon
House in an orchard

I asked myself: why do human beings need to build shelters in order to dwell? The idea which came out from this statement was that this dwelling should be a "non-house" invisible from outside, with no façade nor social status. It should exclusively be experienced from within, with views only towards the sky through light-wells and patios extending the domestic space into the garden.

나는 스스로에게 물었다: 왜 인간들은 거주하기 위해 집을 짓는가? 이 질문을 하면서 이곳에는 입면도 없고 사회적 지위도 없는 외부에서 보이지 않는 "집이 아닌" 주거지가 있어야 한다는 생각을 하게 되었다. 하늘만 보이는 채광창과 내부공간이 정원으로 확장되는 파티오는 전적으로 집 내부에서만 경험되어야 한다.

© Gilles Coulon
House in an orchard

© Gilles Coulon
House in an orchard

The concept of the house could be summarized in one single element: a flat roof preserving the horizontal relation with the orchard trees and the peach-growing walls. The house stretches out at ground level, blending into the garden. The glazed stainless steel box structures reflect the light and help make the transition from interior to exterior. The outline of the house blends into that of the orchard, and is only visible thanks to the horizontal plane of its roof.

이 주택의 컨셉은 과수원의 나무들 그리고 복숭아가 자라는 벽과 수평적 관계를 유지하는 평평한 지붕이라는 하나의 요소로 축약할 수 있다. 대지위에 뻗은 1층 높이는 뜰과 조화를 이룬다. 반짝거리는 스테인리스 스틸 박스 구조물들은 빛을 반사 시키고 집의 내부에서 외부로의 전환을 용이하게 한다. 유일하게 보이는 지붕의 수평면 덕분에 주택과 과수원의 외관이 잘 어우러진다.

### NL Architects

Really hard to choose! We've done many cool roofs. Like the green roof of Blok K in Funenpark that creates a wonderful artificial landscape in an urban setting and gives the sensation of being on vacation.

정말 선택하기 힘들다! 우린 멋진 지붕을 많이 설계했다. 도심속에 아주 멋진 인공 조경을 조성하여 휴가 온 것 같은 느낌을 주는 Funenpark에 위치한 Blok K의 그린루프나 헤

Or the wild roof of the Loop House in Heiry Art Valley. Or the elevated pitch of the BasketBar that offers a public space on top of a small building for shooting hoop. Or the exaggerated profile of Barneveld Central Station. Or even the un-built ones like the sloping parking street ontop of Parkhouse / Carstadt, or the velodrome rooftop of the Bike Pavilion resembling a pagoda or the Skate Bowl of "Leonardo Plaza COO' in Dordrecht with its belly hovering over the interior. The roof is on fire!

The most recent and current favorite: the roof of a club house for three sports clubs in Rotterdam now almost completed. It is a lengthy building placed in between the two main soccer fields and overlooking the hockey pitches. The pitched roof is accessible and serves as a grand stand. So simple : )

이리 마을 Loop House의 와일드한 지붕. BasketBar 작은 건물 꼭대기에서 농구경기를 할 수 있는 개방된 옥상 경기장이나 바네벨트 중앙역처럼 과장된 지붕. 실제로 지어지지는 않았지만, 건물 위에 경사진 노상 주차장이 있는 Pakrhouse (Carstadt)나 경륜장이 있는 Bike Pavilion의 탑을 닮은 지붕. 또는 도르트레히트에 위치한 실내 쪽으로 둥글게 파인 "Leonardo Plaza COO'의 스케이트 경기장 지붕 등 핫 한 지붕은 많다!!

가장 최근 좋아하게 된 지붕은: 현재 거의 완공되어가는 세 종목의 스포츠 클럽이 있는 로트레담에 위치한 클럽하우스 지붕이다. 상당히 긴 건물로 두 개의 메인 축구 경기장 사이에 위치하며 하키 경기장을 바라보고 있다. 경사진 지붕은 그랜드스탠드 형태로 개방된다. 정말 간단하다.

© Raoul Kramer
Verdana

Parkhouse

## object-e architecture

Several of our projects are located in the Greek islands, where the traditional flat roof is enforced by the official regulations. Therefore on the surface of it there aren't many choices concerning the roof. However each of our projects has a different interpretation of the archetypical Cycladic roof and are

우리는 정부 규정하에 전통적인 평평한 지붕이 강요되는 그리스 섬에 여러 프로젝트를 진행하였다. 그래서, 표면적으로는 지붕에 관해서는 선택의 여지가 많지 않았다. 하지만, 각각의 우리 프로젝트는 전통 유형의 건축적 원리에 충실하면서 키클라데스 제도의 전형적

faithful to the architectural principles of the traditional type to a varying degree. The housing prototypes in the islands of the Aegean Sea, along with their roofing system, were developed over a large time span that goes back into the antiquity. While most islands were inhabited already from the 6000 BC, the present form of the towns and the typology of the houses have its roots in the 13th century, when the Byzantine Empire was starting to decline. Several factors affected their urban development and form, including the threat of piracy, the need to have access to the sea, along with the weather conditions and the climate.

Because they were developed, collectively, over such a long period of time, traditional settlements in the islands of Greece have encoded into their 'dna' a kind of collective wisdom; a knowledge accumulated over time. For that they have been studied extensively and they have inspired the work of many modern architects; the case of Le Corbusier being maybe the most famous one. Fumihiko Maki, in his 1964 seminar text "Notes on Collective Form", uses the architecture of the Aegean town as one of his examples that he employs in order to describe what he calls collective form. He implies maybe

인 지붕을 여러 각도로 다르게 해석하고 있다. 지붕 시스템을 포함한 에게 해 주변 섬들의 주거 원형은 과거 고대 시대부터 지금까지 오랜 시간에 걸쳐 발달해왔다. 대부분의 섬에서는 B.C. 6000년부터 이미 인류가 마을 형태를 구축하고 거주하기 시작했으며, 13세기에는 이미 집의 유형이 갖춰졌다. 해적의 위협을 포함해 바다와 근접한 환경, 또 그에 따르는 날씨 조건과 기후 등 여러 요인들이 이들 도시 개발에 영향을 미쳤다. 그렇게 긴 시간동안 집단적으로 발전해왔기 때문에 그리스 섬 지역의 전통 방식은 지역 사람들의 'DNA'속에 인코딩되어 있다. 시간에 걸쳐 축적된 지식. 그렇게 때문에 섬 지역 전통은 광범위하게 연구되었고 많은 현대 건축가들의 작품에 영감을 주었다. 르 꼬르뷔제가 아마 가장 잘 알려진 사례일 것이다. 마키 후미히코는 1964년 그의 세미나 연설 "집단 양식에 대한 기록"에서 집단 양식이라고 하는 것이 무엇인지 설명하기 위해 에게 해 도시 건축을 사례로 들었다. 그는 아마도 지역 건축의 기본 특성 중 하나를 시사했을 것이다. 집과 빌딩은 조경의 오브젝트가 아니라 조경의 일부이다.

이런 맥락에서 섬 지역에 오늘

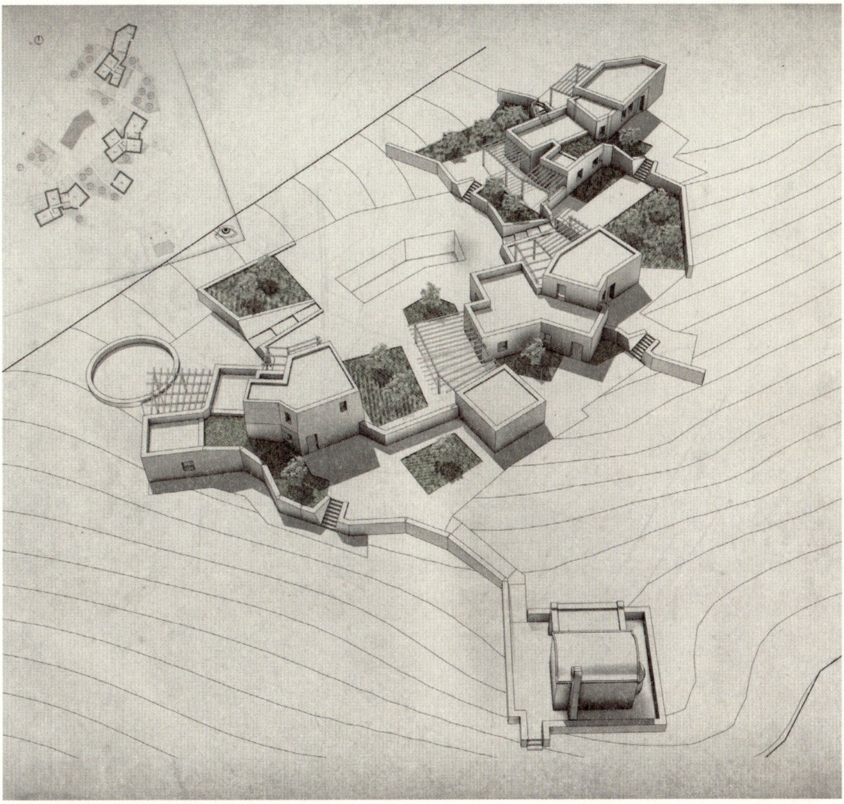

Field Houses

one of the fundamental characteristics of the local architecture: houses and buildings are not objects on the landscape, but instead part of the landscape.

In that context, designing new houses today on those islands is a difficult task. There are in place very strict regulations that are trying to control the growth and protect the traditional architecture. The regulations go at great length in order to specify the requirements, regulating aspects from size, height and relation to the ground, to window types, materials to be used and color. One would think that under those regulations the only option would be to design houses that will look identical to the traditional ones. At the same time, a look at modern development projects on the islands is enough for one to understand that even if the rules are followed faithfully, the result has little to do with the traditional architecture and its fundamental principles; in most cases nothing more than the color (white) and the shapes of the geometry employed (small rectangular volumes). To a certain extend this situation is unavoidable: modern needs, construction methods and requirements are fundamentally different from the ones applied to the traditional housings. At the same

날 새로운 집을 설계하는 것은 힘든 과제다. 이 지역에는 개발을 저지하고 전통적 건축을 보존하고자 하는 차원에서 규제가 매우 엄격하다. 규제 범위는 필요조건들을 명확히 채우기 위해 사이즈 및 높이 제한 그리고 지면과의 관계성, 창문 타입, 재료나 색상 사용 등 상세하게 구체화 되어있다. 이런 규제 하에 집을 설계할 때 선택할 수 있는 유일한 옵션은 전통주거와 동일하게 설계하는 것이라고 생각할지 모른다. 하지만, 이 섬들의 근대 개발 프로젝트를 보고 나면 규제에 충실하면서도 전통 건축이나 기본 원리와는 거리가 먼 결과를 낼 수 있다는 걸 알게 된다. 대부분의 경우 흰색이나 기하학적 형태(작은 직육면체)적용 정도에서 그친다. 불가피하게 현대시대의 니즈와 건설 기술 그리고 요구조건은 전통 주거에 적용된 것들과는 기본적으로 다르다. 그래도 동시에 현재 조건에 매우 도움이 될 만한 전통적 방법들을 찾아 볼 수 있다. 예를 들어 전통 주거는 에너지와 온도 조절 면에서는 매우 효율적이다. 두꺼운 돌 벽은 여름에는 온도를 낮춰주고 겨울에는 따뜻하게 해준다. 작은 창문들은 강한 햇볕을 차단하고, 옥상녹화는 추가적으로 온도 조절을 돕는다. 이런 측면에서 우리가 섬 지역

time though, it is possible to find in the traditional approach elements that could be extremely helpful in the current conditions. For example traditional housing is very efficient in terms of energy and climate control. Thick stone walls allow for cooler temperatures in the summer and wormer in the winter. Small openings provide protection from the intense sunlight. Green roofs provide extra climate control.

In that context, maybe the most successful of our projects on the islands can be found in the 'Field Houses' project. The roofs of the three residencies are of course flat, but are highly differentiated on each of the spaces that they cover: differentiated in shape, size and orientation. At the same time, due to the slop of the ground in many cases the roof of one space is becoming a balcony of another one, while some part are becoming green roofs. That differentiation and micromanaging of the roofs results in something that is maybe very close to the principles, and not only the form, of the traditional Cycladic roofs.

에 설계한 프로젝트 중 가장 성공한 사례는 'Field Houses' 이다. 세 동의 주거 건물의 지붕들은 당연히 평평하다. 하지만, 각 공간을 둘러싸는 외관 디자인이 매우 차별화되었다. 형태와 크기 그리고 방향이 다르다. 또한, 경사진 대지조건으로 인해 한 공간의 지붕이 다른 공간의 발코니가 되며 부분적으로는 녹화 지붕이 된다. 이런 차별화와 지붕의 세세한 처리는 형태뿐 아니라 키클라데스 전통 지붕 원리와도 매우 근접할지 모른다.

## OPARCH

My favorite roof design is Shapeshifter. First, the house is topologically complex, with a large bubble of space passing through the center. The roof bridges over this bubble creating a continuous loop with the adjacent ground. As a result the house feel like a cave carved out of a solid. Second, entire project (including the roof) adheres to the same formal language of simple facets. This unification creates an intentional ambiguity between typically distinct elements like landscape vs. house and walls vs. roof. Also, the faceted geometry creates opportunities for sheared, creased, and pinched forms which encourage dynamic, twisting space. Finally, Shapeshifter really does dramatically change appearance depending on your specific location. This is our first built project to truly explore such parallax effects.

내가 가장 좋아하는 지붕은 Shapeshifter의 지붕이다. 첫째, 커다란 공간 버블이 중앙을 가로지르는 위상적으로 복잡한 주택이다. 지붕의 연속되는 루프형태는 버블 공간위를 가로지르고 땅과 맞닿는다. 그 결과 집은 덩어리를 깎아서 만든 동굴 같은 분위기이다. 둘째, 지붕을 포함한 전체 건물은 서로 동일한 형식 언어를 가진 심플한 양쪽 측면과 맞닿아 있다. 이런 조합은 조경과 집, 벽과 지붕과 같이 일반적으로 대조되는 요소들 사이에 의도된 애매모호함을 창출한다. 또한, 깎인 기하학적 구조는 잘리고, 구겨지고, 죄이는 형태를 만들어내고 역동적이고 구불거리는 공간을 조장한다. 마지막으로, Shapeshifter는 실제로 보는 위치에 따라 겉모습이 급격히 변한다. 이 프로젝트는 이런 시차 효과를 진정으로 실현시킨 우리의 첫 번째 프로젝트이다.

© Hufton + Crow
Shapeshifter

## SLOT STUDIO

The Mexico pavilion for the Shanghai 2010 expo, a commission we were awarded, is our favorite roof design, because it arise from the concept of China's and Mexico's shared cultural features, such as kites, which people play with in both countries. The roof design is an interpretation of this common feature, which serves to articulate the interior and exterior space. Moreover, these kites come in many colors, which is something typical of Mexico.

우리가 설계한 2010 상하이 엑스포를 위한 멕시코관의 지붕이 가장 마음에 든다. 이 지붕은 중국과 멕시코에 공존하는 민속 놀이 중 하나인 연날리기를 컨셉으로 설계되었다. 지붕 디자인은 그런 연의 형상을 반영하며 부스의 실내와 외부 공간을 구분 짓는다. 게다가 색은 멕시코의 정통색상들로 알록달록하다.

Roof Plan

Front Elevation

034 ROOF, protected Open Space

## Stefano Corbo Studio

In my project for a villa in Italy (ABC House), the roof is a complex and irregular geometry that establishes a dialogue with the surrounding orographic context. In other words, the roof is an example of artificial nature: nature is reinterpreted and absorbed into architectural forms.

이탈리아의 ABC 주택에서 나는 주변 산악지형과 닮은 복잡하고 불규칙한 형태의 구조물로 지붕을 설계하였다. 즉, 인공 자연 같다: 재해석된 자연이 건축적 형태로 흡수되었다.

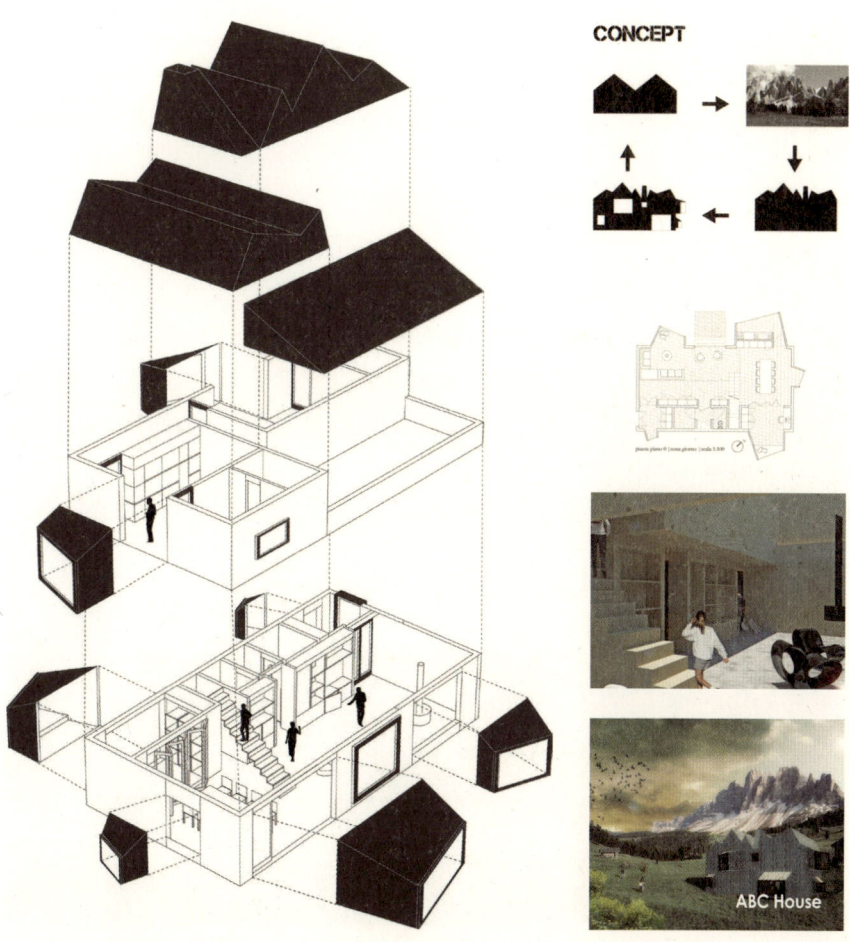

## TOUCH Architect

TREE Sukkasem VILLA, the house on a hillside was designed by emphasizing on sustainability. The roof of this house is one of sustainable elements which has double-layer roof. First layer is made of reinforced concrete with basin-shape for collecting rain water. Second layer is a full panel of wooden trellis above the first layer with a gap between the two layers of roof. This will help avoiding heat from direct sunlight while protecting the roof from heavy rain and damage protection from hailstone.

Tree Sukkasem Villa는 언덕위에 설계된 주택으로 지속가능성에 중점을 두고 설계되었다. 이 주택의 지붕은 이중 레이어 구조로 되어 있어 집의 지속가능한 요소 중 하나이다. 첫 번째 레이어는 철근 콘크리트로 만들어 졌으며 빗물을 모으기 위해 둥근 그릇 형태를 취한다. 두 번째 레이어는 우드 격자 패널들로 덮인 구조로 첫 번째 레이어 위로 살짝 띄워졌다. 이런 레이어 구성은 직사광선의 열을 차단하며, 폭우나 재해로부터 지붕을 안전하게 보호하는데 도움이 된다.

© Chalermwat Wongchompoo

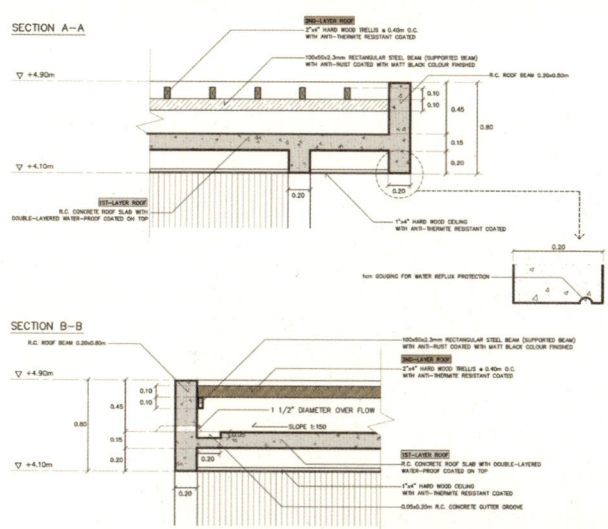

© Chalermwat Wongchompoo

037

TROPICAL SPACE

"OUR FAVORITE ROOF IS IN THE TERRA COTTA STUDIO BECAUSE WE CAN SEE THE SKY IN DIFFERENT WAY, TIME AND EMOTION STATUS."

우리가 가장 좋아하는 지붕은 Terra Cotta Studio의 지붕이다. 왜냐하면 지붕으로 시간에 따라 다양한 모습과 감성이 있는 하늘을 볼 수 있기 때문이다.

© Hiroyuki Oki
terra cotta studio

What is **the least favourite Roof** you've designed and why?

## BOARD

Last year we worked on the design for an extension of the town hall in a small town in Germany called Korbach, where we proposed a one floor high building that connects to all the other new extensions. We placed this connecting building on the first floor to provide occasional access for the public to the centre of the site on the ground floor, where we created a public square with an open-air theatre. We were quite convinced of the success of this elevated building that worked as a roof for the public space underneath. But we got the impression that the prevailing and continuing negative reputation of

작년에 우리는 코르바흐(Korbach)라고 하는 독일의 작은 도시의 시청건물 증축 안 작업을 하였다. 우리는 모든 증축부분과 연결되는 한층 띄어 올려진 건물을 제안하였다. 외부에서 야외 공연장이 있는 건물 중심 바닥 광장안으로 간헐적 접근을 유도하기 위해 이 연결 동을 광장 2층에 배치하였다. 우리는 광장을 위한 지붕으로도 쓰이는 이 띄워진 건물의 성과를 매우 확신하였다. 하지만, 모더니즘이 책망 받는 띄워진 건물 아래 공간에 대한 부정적인 인식은 (언젠가는 극복해야 한다고 믿지만) 여전히 남아있고 만연하다는 느낌

spaces under elevated buildings for which modernism can be blamed - which we believe needs to be overcome - eventually led to the fact that the design was not chosen.

을 받았으며, 이런 우려는 우리 설계안이 채택되지 않는 결과로 이어졌다.

### ELA(Edu Lopez Architects)

In all our projects I am characterized by creating roofs that have a compositive sense with the architecture that is being done, for that reason I am going to show you a project that has a completely flat and functional cover. This does not mean that the cover of this project does not we like, on the contrary, we believe that is a very good solution for the building that was being proposed as a starting point. The building in question is the visitor center of the Reistag in Berlin which we have talked about before. This project, due to its institutional nature and its situation, should reflect a very powerful idea, but at the same time a sobriety that marked the character of it. A volume of

모든 우리 프로젝트에는 복잡하고 개성 있는 지붕이 있다. 하지만, 게 중 가장 평평하고 기능적인 지붕을 소개하겠다. 그렇다고 이 지붕을 좋아하지 않는다는 것이 아니라 반대로 우리는 이 지붕이 설계 해결의 시작점으로써 훌륭했다고 판단한다. 이 건물은 앞에서 언급했던 베를린의 국회의사당이다. 제도적 성향과 환경적 조건 때문에 강렬하지만 냉철한 아이디어로 접근해야 했다. 우리는 다양한 색상 조합이 두드러지는 기학학적 형태의 유리 매스를 제안하였다. 그리고, 그 유리 매스 둘레에 데크(지붕)를 설치하였다. 바깥 기후로부터 건물을 보호하는 역할을 하는 이 데크는 금속으로 만들어졌으며 프로젝트 전체

glass was proposed which followed the geometric shape of the plot, from which a number of colors stood out. On them was placed a deck, which flew from the perimeter of the building. This roof, which was metallic and framed the entire project, is an element that stands on the building and protects it from external climatological elements. This roof has built-in large structure, has a few holes, which make small skylights to introduce light and in other cases to incorporate the buildings own facilities. It is worth noting a very powerful curve on the deck. This curve is due to accommodate and create a space for a tree that should be left and protected.

를 두르고 있어 시각적으로 두드러진다. 데크에 나 있는 작은 구멍들은 내부로 빛을 통과시키며 건물에 필요한 설비시설이 내장되어 있는 거대한 구조물이기도 하다. 나무 한 그루를 그대로 보존하기 위해 매우 극적인 곡선형태를 취하고 있다는 점 또한 주목할 만하다.

The visitor center of the Reistag in Berlin

## LGSMA_

The project is composed by 2 buildings. Structure defines architecture and its language in its purest form. The intention is to employ a material that is widely used in construction in Italy, but trying to turn it into a form of expression, by way of a simple construction system and the use of colour. The achieve is a crude, almost brutal, effect that emphasizes its material quality. Concrete is coloured with red pigment - a color which is very typical for the Roman countryside buildings - in the mixture to get an homogeneous material to mould.

The three pairs of rhomboidal section inclined vertical columns are the building's distinctive features. But I hate the idea of the pitched roof. Starting from the same point on the ground two by two, they open up and fold becoming beams of the roof slab; in-between the pairs of beams, steel cables allow vegetation to grow, creating a protected porch between the main building and the square. These beams outline a geometric grid, characterizing both roof horizontal surfaces, by way of their red color but also through a small intrados relief (6 cm). The roof slab is supported punctually at each beam intersection vertex by steel

(내가 가장 마음에 들지 않는) 프로젝트는 2 개의 건물로 구성된다. 구조적으로는 가장 간결한 형태로 건축 성향을 나타낸다. 이탈리아 시공계에서 흔히 사용하는 재료를 사용하고자 하였다. 하지만, 간단한 시공 시스템과 색상으로 형태적인 변화를 주려고 노력하였다. 투박하고 거친 느낌의 결과물은 재료의 질을 오히려 강조하는 효과를 낳았다. 콘크리트는 거푸집 재료와 동일한 느낌을 주기 위해 붉은 안료로 물들였다. 로마 시골지역 건물에서 볼 수 있는 매우 전형적인 색상이다.

마름모 꼴로 기울어진 3쌍의 수직 기둥은 건물의 독특한 부분이다. 하지만 난 개인적으로 기울어진 지붕을 좋아하지 않는다. 기둥은 쌍쌍이 지상 위 동일한 지점에서 출발하여 위로 갈수록 벌어지다가 지붕 슬라브 빔에서 멈춘다. 각 쌍의 빔 사이에 설치된 강철 케이블들은 본관과 광장 사이에 중문 형태의 입구를 형성하며, 케이블을 따라 식물들이 자란다. 이 빔들의 윤곽은 그리드 형태를 띠며 붉은 색상과 작은 내륜 릴리프(6cm) 때문에 양쪽 지붕의 수평면이 부각된다. 지붕 슬라브는 모든 빔 교차점에서 강철 기둥으로 일정하게 지

columns.
Outer public and semi-private spaces adjoining the building, as well as its interior, are conceived as an extension of domestic space, hosting the spontaneous social initiatives of the inhabitants. As in the local practice of use of land, countryside and urbanity merge together, for an ecologic conversion of territory into a living landscape.
The roof of the second building has been thought as an extention of the public space; it is a square.
Each component of the building for Ian+ is an occasion of design.

탱된다.
건물 그리고 건물 내부와 접해 있는 외부 공공 공간과 준 사적 공간은 건물 내부가 확장된 것처럼 보이며 이웃 주민들의 자발적인 교류가 이뤄지는 곳이다. 토지 활용 지역 관행에 의하면 토지를 생태적인 생활 환경으로 전환하기 위해 시골과 도시가 하나로 병합되는 추세이다.
별관의 지붕은 광장처럼 공공 공간의 연장으로 쓰인다.
Ian+에게 모든 건물의 구성요소는 디자인의 기회이다.

### Miha Volgemut architect

All the flat roofs I designed are of no significance, because the flat roof is not a roof at all.

내가 설계한 모든 특이성 없는 평평한 지붕들이다. 평평한 지붕은 지붕이 아니다.

### Moussafir Architectes

Flat roofs with solar panels made to fit politically correct urban regulations.

정치적인 도시 규제에 맞춘 태양열 판넬이 부착된 평평한 지붕들이다.

### NL Architects

I can't think of one right now! But sometimes reality is counter intuitive. We recently completed a very nice school in Knokke Heist in Belgium. We wanted all the benefits of a vegetation roof and we intended to collect rain water for flushing toilets and so on. Paradoxically we had to reconsider the sedum and install pebbles on a large stretch of the roof: surprisingly the grey roof was greener than a green roof...

지금 당장 생각나는 건 없다! 하지만, 가끔씩 현실은 직감과 반대라는 걸 느낀다. 최근 우리는 벨기에 크노케 하이스트에 매우 멋진 학교 건물을 완공하였다. 우리는 수세식 화장실에 활용하기 위해 빗물을 모으는 등 친환경 지붕이 가진 모든 이점을 누리고 싶었지만, 역설적으로 우리는 돌나무 대신 지붕의 많은 부분에 자갈을 깔아야 했다: 놀랍게도 이 회색 빛 지붕은 그린루프보다 더 녹색이었다...

© Marcel van der Burg
School in Knokke Heist in Belgium

## OPARCH

Any roof made of Siplast or the equivalent. We are at three and counting. These roofs are technically great, but the typical details are intended to be hidden. It is pretty difficult to have much emotional attachment to a roof concealed behind a parapet and never seen.

Siplast나 비슷한 재료로 만들어진 지붕이다. 이런 지붕을 세 번 디자인 해봤고, 우린 그걸 인정한다. 이런 지붕은 기술적으로는 훌륭하지만, 대부분의 디테일이 숨겨진다. 난간 뒤에 숨겨져 전혀 보이지 않는 지붕에는 애착을 가지기가 매우 힘들다.

## SLOT STUDIO

In 2009 we participated in a tender issued by the Mayor's Office of Amsterdam to develop a residential and office complex in an underused industrial zone of the city. Slot Studio views each design commission as an opportunity to explore; even though on this occasion, the major thrust of the design direction focused on the necessary urban presence of the building in terms of volumetrics, and the formal expression of facades and roof structures, on a scale not often explored.

2009년 암스트레담의 시청사가 주최한 설계입찰 공모에 참여하였다. 도시의 유휴산업지역에 주상복합건물을 짓기 위한 공모였고, 우리에게 모든 디자인 지침들은 연구할 수 있는 기회로 보였다. 여기에서도 우리의 주된 디자인 취지는 규모 면으로나 입면의 형식적 표현 및 지붕구조 면에서 이 흔히 접할 수 없는 스케일을 가진 건물의 도시적 존재감을 부각시키는데 초점이 맞춰졌다.

1. Toilets
2. Lift
3. Offices
4. Meeting Rooms
5. Cubicles
6. Terrace

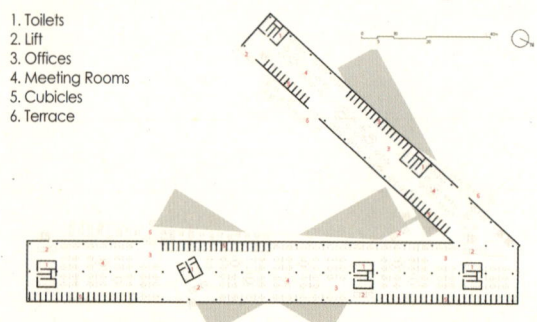

First Floor

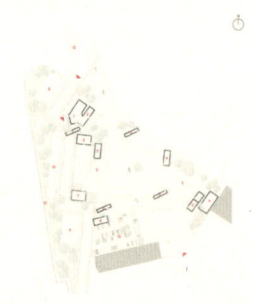

Ground Floor

©Tim Griffith
Hyde in USA

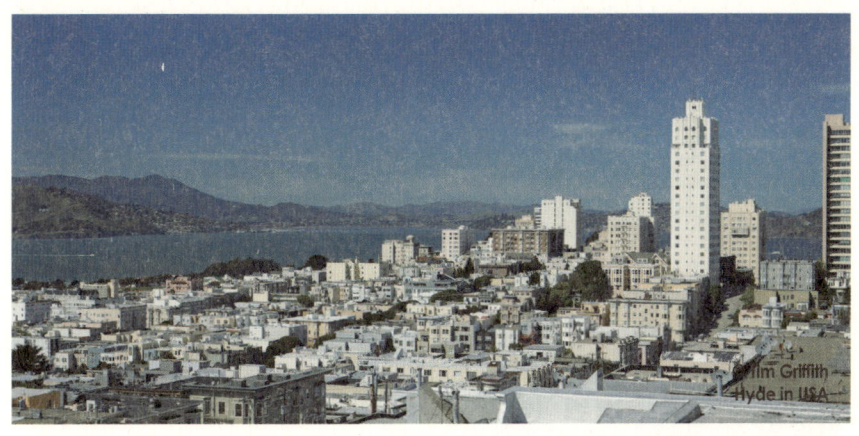

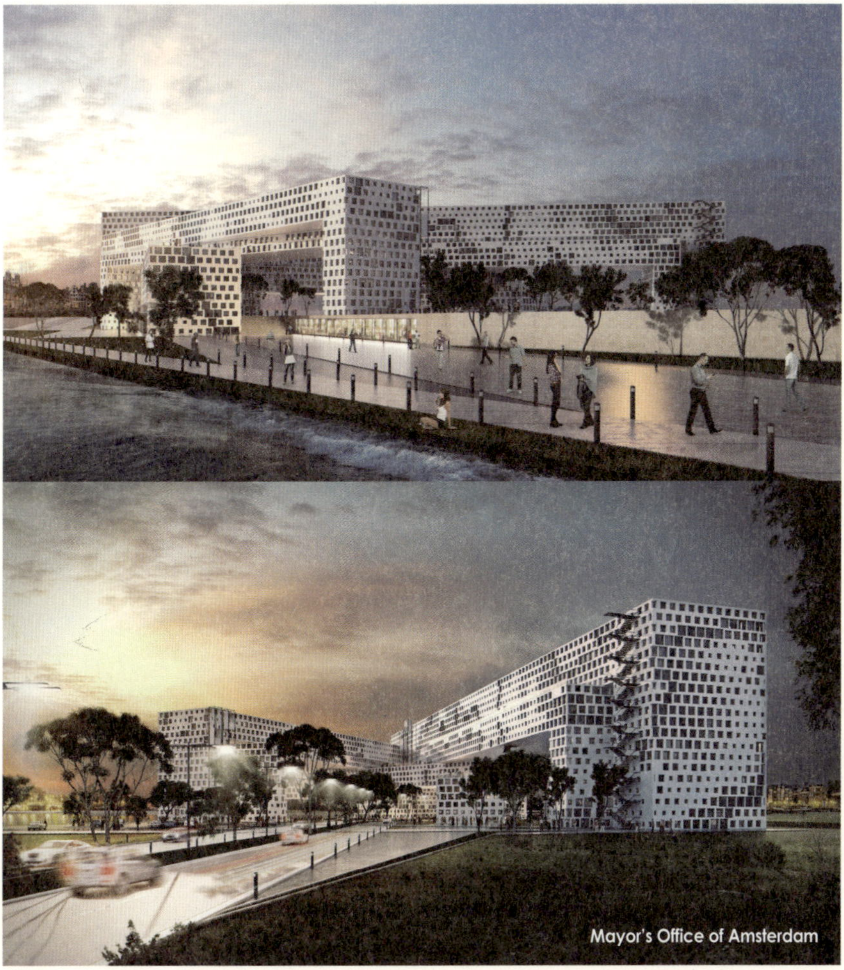

Mayor's Office of Amsterdam

## Stefano Corbo Studio

In my proposal for the Industrial Arts Centre in Cincinnati (an example of adaptive reuse), the roof is just an empty surface: it's neither a green roof, nor a panoramic terrace. The main design efforts were addressed to the interior of the building, and at the end the roof looked like an incomplete part of the project.

신시내티의 기술 센터를 위한 (적응적 재활용의 본보기로) 제안한 설계안의 지붕으로 그저 텅 빈 외피이다: 녹색지붕도 경치 좋은 테라스도 아니다. 지붕의 내부 디자인에 가장 공을 많이 들였으며, 지붕 가장자리는 마치 미완성된 것처럼 보인다.

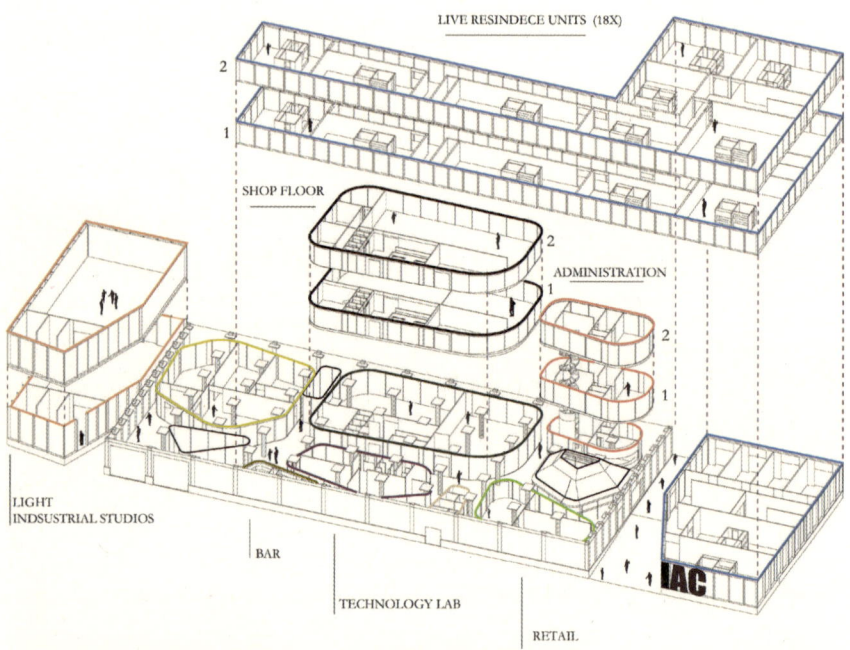

050 ROOF, protected Open Space

# TROPICAL SPACE

Also the roof of terra cotta studio. We think without the glass, it will be much more interesting.

가장 좋아하지 않는 지붕은 Terra Cotta Studio의 지붕이다. 유리가 없었다면 더 매력적인 지붕이 되었을 거다.

# Q3

What is
**the most
typical type and
color of roofs**
in your country?
Is there **a reason for that?**

# FRANCE
## Moussafir Architectes

Traditionally, they are clay red in the south of France (Terracotta roof tiles), dark grey in the north (zinc or slates) or with bright colors (glazed tiles) in Burgundy.

전통적으로 프랑스 남부지역은 붉은 진흙 색(테라코타 지붕 타일)이며, 북부지역은 진한 회색(아연판이나 슬레이트)이나 밝은 버건디 계열(채유타일)이다.

# GREECE
## object-e architecture

If we were to distinguish just two archetypical roof types in the traditional architecture of Greece, we would have to follow a geographical approach: On the one hand we have the roof type that we encounter in the mainland: Pitched roofs, most of the times gabled but also hipped ones. Especially in the most mountainous villages and in the northern part of the country they define the rule in traditional architecture. Most of the times the covering material is ceramic tiles, generating the characteristic earthy red color. However, stone can also be used. The second type of roof can be found on the islands of the Aegean sea: Flat roofs at different levels for each of the spaces of the house that they are covering. Those roofs were traditionally made out of straws, seaweed, red earth and clay soil. However in many cases the top is painted white, defining therefore their characteristic color.

그리스 전통 건축의 지붕 원형 타입을 단지 두 종류로 분류한다면 우리는 지리학적으로 구분할 것이다. 그 중 하나로 그리스 본토에서 볼 수 있는 지붕타입이 있다. 경사진 지붕. 대부분 박공지붕이지만 모임지붕도 있다. 특히, 대부분의 산지촌이나 북쪽 지방에서는 전통 건축을 따를 것을 규정하고 있다. 마감재로는 대부분 진흙의 자연 붉은 색이 도는 세라믹 타일을 사용하지만, 돌을 사용하기도 한다. 두 번째 지붕타입으로는 에게 해 주변 섬 지방에서 볼 수 있는 지붕으로 집의 공간 구성에 따라 높이가 다른 평평한 지붕이다. 이러한 지붕은 전통적으로 짚이나 해초, 또는 붉은 진흙이나 식토로 지어진다. 그러나, 많은 경우 흰색으로 도색 되어 흰색이 이 지역 지붕 특유의 색으로 인식되었다.

Typical Cycladic roofs – the city of Paros

Typical Cycladic roofs – the old city of Kavala

In modern Greek cities however, we are encountering a different type of roof that is prominent: Flat concrete roofs that one could argue that they are the result of the influence of modern architecture. However, they ways in which modern architecture principles were applied in Greek cities are rather unique and manage to create, unintentionally, a new type. Modern Greek cities were developed in the 1960s through a unique system where the owner of the lot would allow a constructor to build a typically 4 to 6 stories apartment building and in return the owner would get two or three of the apartments. The constructor would go on to sell the rest of the apartments. The result of that process was relatively small buildings, highly unregulated, that would follow more or less the same architectural principles: concrete structures with flat roofs, small balconies and usually

그리스 현대 도시에서는 차별된 유형의 두드러지는 지붕들을 볼 수 있다. 현대 건축의 영향을 받았다고 주장할 할 만한 평평한 콘크리트 지붕이다. 하지만, 그리스 도시에는 현대적 건축 이론이 독특하고도 의도치 않은 새로운 방향으로 적용되었다. 그리스 현대 도시는 1960년대 개발되었으며 대지 소유주가 건설사에게 일반적으로 4층에서 6층 높이의 아파트 빌딩만을 건설하도록 허용하는 독특한 시스템이 적용되었다. 대신 대지 소유주는 두 채 내지 세 채의 집을 소유할 수 있었고, 건설사들은 그 나머지 집들을 분양하였다. 이런 과정을 거치다 보니 평평한 지붕과 좁은 발코니 그리고 거

Roof-scapes of the modern Greek City – the city of Thessaloniki.

lack of any decorative or traditional elements. Colors are also rare, with white and grey being the rule.

의 장식이나 전통적 요소가 없는 콘크리트 건물과 같은, 거의 동일한 건축적 이론을 따르는 정돈되지 않고 상대적으로 작은 건물들이 생겨났다.

The modern Greek City – the city of Thessaloniki. On the left front side are the pitched roofs of the old city, in contrast with the 'new typology' of the modern city.

Roof-scapes of the modern Greek City – the city of Thessaloniki.

Those buildings, made in many occasions without any concern for architectural quality are generally bad designs that result in really ugly buildings. Surprisingly enough however, through their immerse repetition they create a unique cityscape that has some emergent qualities. The roofs are one of the most characteristic aspect: The flat, neutral roofs are covered by an endless number of TV antennas (one for each apartment), ad hoc constructions added during the life of the building, and more recently solar panels. The flatness of the roof is occupied by a chaotic but in many ways interesting aggregation of vertical elements. Therefore the 'modern Greek roof' becomes an unintentional new type with its own architectural characteristics.

색상 또한 흰색과 회색으로 단순하게 규정되어 버렸다. 상당한 경우가 어떠한 건축적 가치를 고려하지 않았으며, 대부분 질 낮은 디자인의 매우 형편없는 건물들이었다. 하지만, 놀랍게도 이런 건물들이 집단적으로 반복되면서 독특한 도시경관을 형성하며 나름 신흥적 가치를 발산한다. 그 중 평평하고 단순한 지붕 위를 TV 안테나(집집 마다 하나씩)가 끝없이 뒤덮고 있는 모습은 가장 특성 있는 부분 중 하나이다. 사람들이 거주하면서 그때 그때 설치한 것들이다. 최근에는 태양 전지판이 더 많다. 지붕의 평평함은 복잡함으로 가득 차 있다. 그러나 여러모로 수직적 요소들의 흥미로운 집합체로 보인다. 결론적으로, '현대 그리스 지붕'은 의도치 않게 자신만의 건축적 특성을 가지게 된 새로운 유형으로 자리 잡았다.

Roof-scape of the modern Greek City - details

The skyline of the modern Greek City defined by all the add-ons of the rooftops.

# ITALY
## b4 architects

In Italy the traditional roof in all historical cities is made with imbrices and flat tiles, the most part of them are made in terracotta, so in general the most diffuse colour on the roofscape in the cities is red/brown in different shade depending of the chemical quantity and type of clay inside the composition. In these last decades however are catching on all the modern industrial roof covering used in the contemporary architecture in a very wide of range of materials. There are some regions that conserve some traditional way to build roof strongly

이탈리아 모든 역사적 도시의 전통 지붕은 수키(볼록한 타일)와 평평한 타일로 만들어졌으며, 대부분 재료가 테라코타이다. 그래서 가장 많이 보이는 지붕 경관 색은 적/갈색으로 지붕에 쓰인 화학재료의 함유량과 점토의 종류에 따라 다양한 형태를 취한다. 반면 최근 몇 세기동안 유행하는 현대적 건축디자인의 산업화된 지붕에는 광범위한 종류의 재료가 사용된다. 인근 지역 중 전통방식을 고수하여 자연 재료만으로 지붕을 짓는 곳들이 있다. 예를 들어, 제노바 주변 리구리아에서는 슬레이트에

depending of natural resources on local surroundings, like special stones for example slate in Liguria, around Genoa with a dark grey colour.

진 회색의 특이한 돌만 사용한다.

"IN ITALY THE TRADITIONAL ROOF IN ALL HISTORICAL CITIES IS MADE WITH IMBRICES AND FLAT TILES, THE MOST PART OF THEM ARE MADE IN TERRACOTTA."

© G. Evels
Piancastagnaio

# ITALY
## Stefano Corbo Studio

Italy has a double significant vernacular tradition: in the northern areas of the country, roofs are made of terracotta and clay tiles. On the contrary, in the south, the Mediterranean architectural legacy is still strong: traditional buildings can be white and their roofs almost horizontal.

이탈리아의 일반 전통 주택에는 이중적 의미가 있다: 북쪽 지방에서는 테라코타와 점토 타일로 지붕을 만드는 반면 남쪽 지방에는 지중해식 건축 유산이 여전히 강하게 남아있다: 전통적 건물은 흰색일 경우가 많으며 지붕은 대부분 평평하다.

# JAPAN
## Katsuhiro Miyamoto & Associates

Typical roof colours are grey which is due to the commonly used clay roof tiles. However one can observe red to brownish tones mainly in the western part of Japan.

전형적인 지붕 색은 회색으로 통상적으로 사용되는 기와 색이다. 하지만, 주로 일본 서부 지역에는 붉거나 황토색도 눈에 띈다.

© Artem Canli
Port of Corricella in Procida

## MEXICO
### Ezequiel Farca + Cristina Grappin

In some areas of the center of the country the floors of the terraces are covered with slabs of clay, a functional solution that undoubtedly establishes a traditional chromatic where earth tones predominate and coexist satisfactorily with the green of potted plants and some accents of colors in tile and walls. Undoubtedly it is the traditional image of the Mexican open space.

우리나라 중앙 어느 지역에는 테라스 지붕이 점토타일로 덮여 있다. 점토타일은 기능중심 해결방안으로 두드러지는 진흙 톤을 띤 전통색은 초록 화분과 강한 엑센트 색의 타일이나 벽과 꽤 잘 어울린다. 이런 조화는 멕시코 공용공간의 전통적인 이미지이다.

## MEXICO
### SLOT STUDIO

Most roof exteriors are flat and finished in a brick red color, which is the pigment of the most common impermeable roofing finish products.

대부분의 지붕이 평평하고 방수가 되는 가장 흔한 지붕 마감재인 붉은 벽돌로 마감되었다.

## NETHERLANDS
### BOARD

When you look at Rotterdam from above and with the help of Google Maps, you will see a lot of grey roofs, which has to do with the large number of modern buildings with flat roofs in the city. Most of these buildings were built after WWII when the centre of Rotterdam was almost

구글 맵을 통해 상공에서 로테르담을 보면 회색 지붕이 많이 보일 것이다. 이는 도시에 평평한 지붕으로 된 현대적 건물이 많기 때문이다. 대부분의 이런 건물들은 로테르담 중심부가 거의 파괴되었던 세계 2차대전 이후에 지어졌다. 회색

© Carlos Delgado
Aerial view of Madrid

entirely destroyed. The grey colour comes typically from the bitumen layer or stone chippings on the flat roofs. Nevertheless, you will find a certain number of pitched roofs from the prewar era too, which normally are cladded with roof tiles that come in different colours such as red, brown, but also grey. Since cities such as Amsterdam were less affected during WWII, the number of pitched roofs there is higher which makes these cities appear more reddish when viewed from above. However, the prevailing colour there is rather grey too.

은 대개 평평한 지붕위에 깔린 아스팔트나 조각 돌의 색이다. 그 와중에 전쟁 이전 시대의 붉은색, 황토색, 회색 등 다양한 색상으로 된 일반적인 기와 경사지붕도 어느정도 볼 수 있다. 암스테르담과 같은 도시들은 세계2차대전의 영향을 덜 받은 곳이기 때문에 경사 지붕의 수가 더 많아 상공에서 봤을 때 도시가 좀 더 붉은색으로 보인다. 하지만, 가장 압도적인 색은 여전히 회색이다.

## NETHERLANDS
### NL Architects

Dutch roof tiles are often red. The river clay turns red in the oven.

네덜란드 지붕 기와는 대부분 붉은색이다. 강 속 진흙을 화로에 구우면 붉게 변한다.

## SLOVENIA
### Miha Volgemut architect

In my country the most characteristic colour of the roof is probably brick-orange, that prevails in old medieval towns across Slovenia. The reason is probably that the clay was always the most permanent material in the past.

우리나라에서 가장 흔한 지붕은 아마도 슬로베니아 전역 오래된 중세 도시에 만연한 주황 벽돌 색일 것이다. 그 이유는 과거에는 점토가 늘 가장 영구적인 자원이었기 때문일 것이다.

© Tim Griffith

## SPAIN
### ELA(Edu Lopez Architects)

There are many types of roofs in Spain. At present, all the buildings that are made, mainly residential, are covered flat and passable with a base of white gravel, for reasons that usually these covers are usually used to put air conditioning installations or to put clothes lines. Other types of covers are those made before that the flat ones and have an orange or reddish appearance. This color is due to the very material of the covers that are

스페인은 지붕의 종류가 다양하다. 현재는 주거건물이 대부분으로 모든 주거건물의 옥상은 평평하며 흰 자갈이 깔려있다. 그래야만 에어컨 실외기나 빨랫줄을 둘 수 있기 때문이다. 이런 평평한 옥상 이전의 지붕은 오렌지색이나 붉은색이었다. 스페인은 전통적으로 세라믹과 벽돌을 사용하며 이러한 색은 세라믹 원료에서 나온 것이다. 그렇기 때문에 지붕들이 매우 개성이 넘친다.

ceramic. Spain has a great tradition in the ceramics and brick and for that reason they are created of this material that gives that that aspect so characteristic.

## USA
## OPARCH

In California white or a light metal is typical, to avoid the heat island effect. Green roofs have also become increasingly common as the cost for the requisite assemblies has become more competitive.

캘리포니아에서는 열섬 효과를 피하기위해 흰색 이나 경금속이 일반적으로 쓰인다. 녹색 지붕 또한 설치비용이 예전보다 저렴해 져서 갈수록 흔해진다.

# THAILAND
## TOUCH Architect

Red-orange-brown tone of roof is the most typical colour of gable and hip roof in Thailand, since it derived from local material which is an earthenware. The earthenware tile has a combination of red, orange, and brown colour, since it was made of baked clay. The shape of the roof is also important which has to be a steep slope roof because of heavy rain in Thailand. It leads through the material using that an earthenware tile is the most suitable and favourable for the roof material.

지붕의 빨강-주황-황토 색조는 현지 도기 재료에서 유래된 색들로 태국의 박공 및 너새 지붕에서 볼 수 있는 전형적인 색 조합이다. 도기 타일의 색이 빨강-주황-황토인 이유는 구운 점토로 만들었기 때문이다. 지붕 형태 또한 중요하다. 태국에는 폭우가 많이 내리기 때문에 지붕 형태가 가파른 경사 모양이어야 한다. 이런 지붕 형태에는 도기 타일이 가장 적합하고 좋다.

© Setthakarn Yangderm

"THE TILES MAYBE THE MOST TYPICAL ROOF IN VIETNAM BECAUSE OF THE TROPICAL CLIMATE WITH LOTS OF RAIN AND THE HOT SUMMER."

# VIETNAM
# TROPICAL SPACE

비가 많이 내리고 더운 열대기후 때문에 베트남의 가장 전형적인 지붕은 아마도 기와 지붕일 거다.

© Diane Selwyn

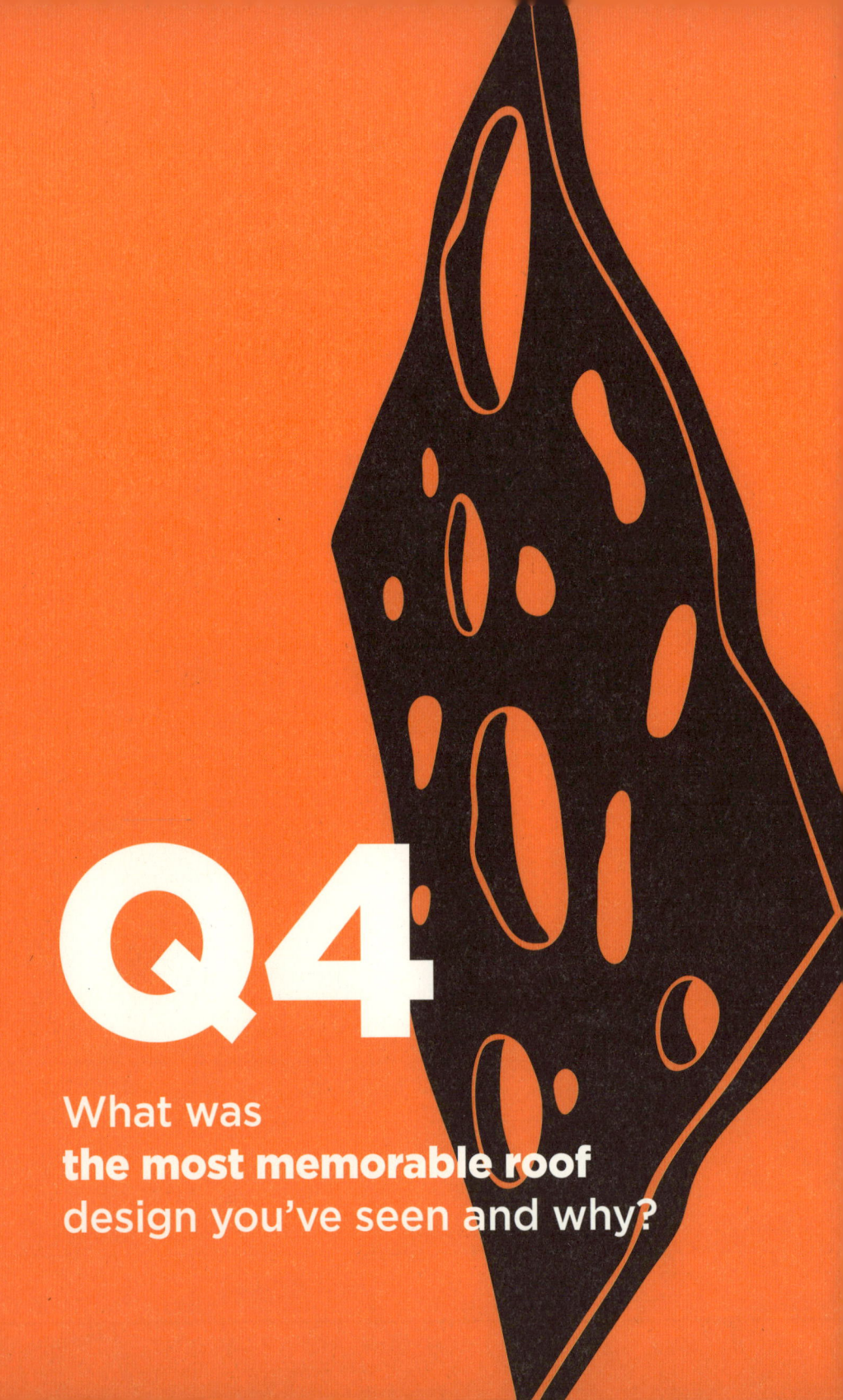

# Q4

What was **the most memorable roof** design you've seen and why?

## b4 architects

The great culture and congress center designed by Jean Nouvel in Lucerna Switzerland, has an impressive roof that is the synthesis of a complex project.

장 누벨이 디자인한 스위스의 루체른의 거대한 문화학술센터에는 복합물과 조화를 이루는 인상적인 지붕이 있다.

© G. Evels
Culture and Congress Centre

Culture and Congress Center
Lucerna - Switzerland - Nouvel

The brilliant structural solution and the design expedient to finish the roof as a sheet of paper gives to the roof an apparent lightness in great contrast with the dimension of the cantilevered structure. The special location on the lake with all its reflecting effect, especially in the night make this work an icon of the last contemporary architecture.

Here in Rome we love the great work of Pier Luigi Nervi and his experimental sense of space applied to great structural concrete architectural events, like the Paul VI Audience Hall in Vatican City: the natural lighting incoming from the prefab roof gives a special lightness to the great vault.

한 장의 종이와 같은 지붕을 완성하기 위해 처방된 똑똑한 구조적 해법과 디자인은 캔틸레버 구조의 큰 지붕과는 대조되는 가벼움을 선사한다. 호수 위라는 장소적 특이성은 물에 비치는 반사효과를 내며 특히 밤의 효과는 현대 건축의 끝장 아이콘으로 손꼽힌다. 우리는 이곳 로마의 피에르 루이지 네르비의 작품을 매우 좋아한다. 그의 실험적인 공간적 감각은 바티칸의 바오로 6세 알현실과 같이 훌륭한 콘크리트 구조 건축물에 적용되었다: 조립식 지붕에서 들어오는 자연광은 거대한 아치 천장이 가벼워 보이는 특수한 효과를 준다.

© S.Papitto
Paul VI Audience Hall

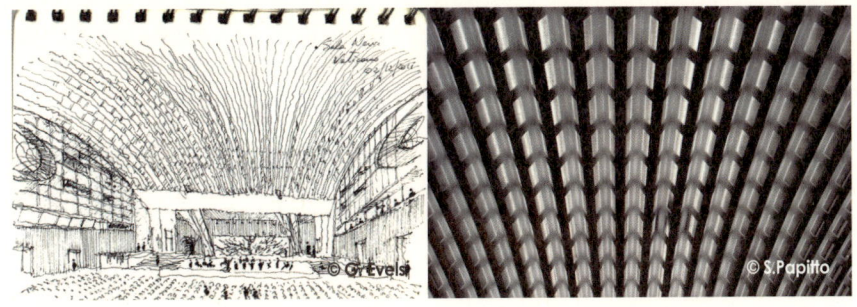

## BOARD

One of the roof structures that I consider as outstanding is the São Paulo Museum of Art in Brazil designed by Lina Bo Bardi. What I am referring to in particular is the space under the building, where the elevated museum in its entirety becomes the roof of that space. To elevate a public building, which is usually supposed to have a strong presence on the ground, in such a way I consider as very brave and fascinating, especially as this open space becomes a place for the people that can be accessed free of charge and used flexibly for many things such as markets, performances, or other events.

두드러지게 뛰어나다고 생각하는 지붕 구조는 리나 보 바르디가 설계한 브라질의 상파울루 미술관이다. 특히 내가 주목하는 점은 건물 아래 공간이다. 전체적으로 띄워진 박물관은 아래 공간의 지붕이 된다. 보통 지상높이에서도 강한 존재감을 드러내는 공공 건물을 띄운다는 것은 매우 용감하고 감탄할 만한 결정이라고 생각한다. 특히, 통행료 없이 자유롭게 지나다닐 수 있는 개방된 공간이 생긴다는 것은 마켓이나 공연 혹은 다른 여러 행사와 활동을 위해 자유롭게 활용할 수 있는 기회가 생긴다는 의미이기도 하다.

© Matt Kieffer
São Paulo Museum of Art

### ELA(Edu Lopez Architects)

One of the covers that most impressed me when I saw them for the first time, is certainly the cover of terminal 4 of the Madrid airport created by Richard Rogers and Estudio Lamela. The legible and modular design of the building creates a repetitive sequence of waves formed by huge prefabricated steel wings. It is based on central structural "trees", of double wingspan. The wooden roof is marked by the "guns" that allow the entrance of natural light that crosses the open spaces of the different plants. Inside the striking metallic corrugated structure of the roof is lined with bamboo strips that give it a smooth and simple appearance, creating a gallery of 1.2 kilometers long.

처음 본 순간 감동받은 지붕은 Richard Rogers와 Estudio Lamela가 가 디자인한 마드리드 공항 터미널 4의 지붕이다. 눈에 확 띄고 모듈화된 디자인은 거대한 조립식 강철 날개 구조가 파도처럼 반복된다. 양쪽 날개는 가운데 "줄기" 구조로 지탱되며, "총알 구멍"이 뚫린 목재 지붕 틈으로 자연광이 비춘다. 커다란 금속 골 구조안에서 대나무 줄기들이 뻗어 나와 부드럽고 심플한 지붕이 되고 이 지붕 밑으로 1.2km 길이의 갤러리가 형성된다.

© Diego Delso
Terminal 4 of the Madrid airport

The striking skylights that pierce the roof have white interior brise-solei. On the outside the roof was finished in aluminum. Undoubtedly the wave of the deck is evident both in the interior and in the experience of the project, as we usually do with our projects. It is really spectacular to see 1.2 km of length of said deck.

지붕의 브리이즈솔레일(햇볕을 가리기 위해 건물의 차에 댄 차양)을 통해 강한 채광이 지붕을 뚫고 들어온다. 지붕 바깥 부분은 알루미늄으로 마감되었다. 우리 프로젝트도 그러듯 터미널 4 지붕의 웨이브도 건물 내부에서나 외부에서나 확실하게 느낄 수 있다. 이런 지붕 아래에 조성된 1.2km 길이의 공간은 그야말로 장관이다.

### Ezequiel Farca + Cristina Grappin

The Terrace of House-Studio of Diego Rivera and Frida like me enough for its history and what they represent as an artistic discourse. If we look closely at these roofs (rooftops) we will observe slabs of clay, like the traditional houses

프리다와 디에고의 House-Studio의 테라스는 역사적으로나 표현된 예술적 담론면으로나 충분히 우리를 만족시킨다. 지붕을 자세히 보면 앞서 언급한 전통 주택에서 볼 수 있는 점토타일이 보인다.

© Claudia Beatriz Aguilar
Diego Rivera and Frida Kahlo Studio

mentioned before, which tells us about the functionalist movement in Mexico, which take back traditional elements and combine them with the formal and the principles of the international style. It seems to me a strong proposal, that is a great reference today. Speaking of the use, today we can see them as an open area, full of possibilities and with a natural context (many trees) that shelters the use that you want to give.

전통적인 요소로 되돌아가 인터내셔널 스타일 형식 및 사상과 결합하는 멕시코 기능주의적 움직임이 엿보인다.  이 지붕은 우리에게 거부 할 수 없는 제안으로 요즘 훌륭한 참고 자료로 쓰인다. 지붕의 사용성에 대해 말하자면, 오늘날의 지붕은 주변 자연(많은 나무)의 보호아래 무한한 가능성으로 채워진 열린 공간이라고 볼 수 있다.

### Katsuhiro Miyamoto & Associates

The steel roll factory in Kitakyushu recently uncovered to be designed by Togo Murano has a roof form that reminds me of three cats with lovely ears.

최근 토고 마라노가 설계한 것으로 밝혀진 기타슈큐에 있는 스틸롤 공장의 지붕은 귀여운 귀를 가진 세 마리의 고양이를 연상시킨다.

© Katsuhiro Miyamoto & Associates
The steel roll factory

## Miha Volgemut architect

It was probably Coop Himmelb(l)au's rooftop at Falkestrasse in Vienna that changed my view of architecture; the freedom, courage and will power that architecture can embody.

비엔나에 있는 쿱 힘멜블라우가 설계한 팔레스트라스 감옥의 지붕으로 이 지붕은 건축의 가능성과 잠재력에 대한 나의 관점에 변화를 주었다; 건축이 구현할 수 있는 자유, 용기 그리고 의지력이 담겨있다.

Coop Himmelb(l)au's rooftop sketch by Miha Volgemut architect architect

Moussafir Architectes

The roof of the Château de Chambord in the Loire valley always fascinated me with its expressive chimneys, huge roof windows and towers forming a refined and artificial landscape on top of a massive medieval base. These objects inserted in the high pitched roofs seem disconnected and out of scale. The whole roof appears like a fairy tale setting raised above ground level.

루아르 계곡에 있는 샹보르 성 위 강렬한 굴뚝과 커다란 지붕 창 그리고 거대한 중세 건축물 위로 세련된 인공 조경을 형성하는 타워에 나는 늘 매혹된다. 높이 치솟은 지붕에 삽입된 이 요소들은 균형을 잃은 채 서로 떨어져 있는 것처럼 보인다. 전체 지붕의 모습은 지상위로 떠오른 동화 속 한 장면 같다.

# NL Architects

Azuma House by Tadao Ando is mind blowing. It has no roof. The house is organized around a small courtyard. You have to cross this outdoor 'room' if you want to go from the kitchen to the living room or from the sleeping room to the toilet.

There is this fantastic rule based home by Kazuo Shinohara called House Under High Voltage Lines. It is based on the regulation that Power lines have a radius around them that you're not allowed to inhabit. Shinohara simply subtracted this extruded radius form an archetypal house, resulting in a mighty strange exterior and magnificent interior.

안도 다다오의 아즈마 하우스가 감동적이다. 이 주택에는 지붕이 없다. 작은 중정 주변으로 공간이 구성되었는데, 부엌에서 거실로 이동하거나 침실에서 화장실로 이동하려면 이 '옥외 방'을 지나가야 한다. '고압선 밑 주택'이라고 하는 시노하라 카즈오가 규정한 기가 막힌 주택론이 있다. 고압선 반경안에는 거주할 수 없다 라는 개념을 바탕으로 시노하라는 고압선에 의해 돌출되는 반경을 주택 원형에서 단순히 깎아 내 버림으로써 결과적으로 매우 특이한 외관과 매우 아름다운 내부를 만들어 냈다.

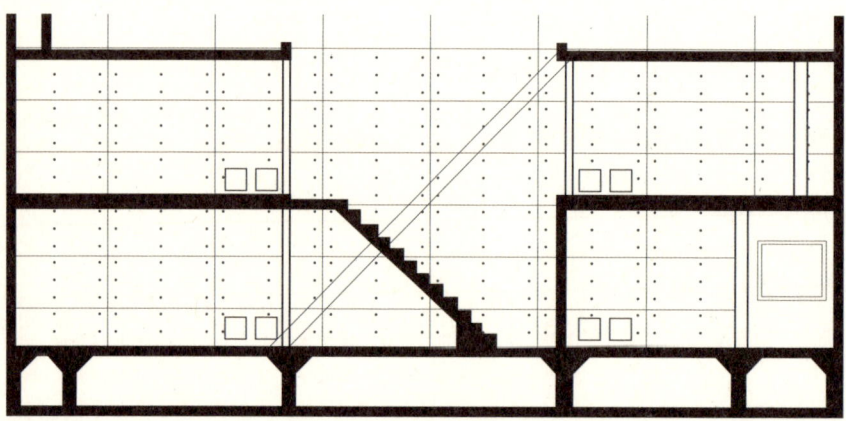

Azuma house section

This summer we were staying a few days in the Unité d'Habitation by Corbu in Marseille. In the evening the residents go up to watch the stars. Or eat pizza. Or dance to a deejay if there is a block party. Children run around like on the village square, they play in the pool, or climb rocks that resemble the mountains in the distance. Heaven.

이번 여름 우리는 마르세유에 있는 르 꼬르뷔제의 유니테 다 비타시옹에서 며칠 지냈다. 저녁이 되면 주민들은 별을 보기 위해 옥상위로 올라간다. 주민 파티가 있는 경우에는 피자를 먹거나 디제이가 트는 음악에 맞춰 춤을 추기 위해 올라가기도 한다. 어린이들은 마을 광장에서 뛰어 다니거나 수영장에서 수영을 하거나 저 멀리 산처럼 솟아오른 바위 들을 오르기도 한다. 그곳은 천국이었다.

Shinohara House

Unité d'habitation

## OPARCH

The most memorable roof design I have seen is the Rolex Learning Center by Sanaa in Lausanne, Switzerland. I particularly like entering under the floor, which doubles as a roof for the various outdoor spaces. This doubling is very effective and heightens the contrast between the interior topography and the flat site. It is an odd building to visit, because it is so vast and relentless. After an hour or so you start to forget the world you left behind and begin to wonder why the entire world isn't built with gentle curves and minimal details.

우리가 본 것 중 가장 인상깊은 지붕은 SANAA가 디자인한 스위스 로잔에 있는 로렉스 러닝 센터이다. 특히 지붕과 겹쳐진 모양으로 여러 외부공간의 지붕 역할을 하는 바닥 하부 밑으로 진입하는 것을 좋아한다. 이런 겹쳐짐은 매우 효율적이며 실내 지형과 평평한 대지사이의 대비효과를 상승시킨다. 이 건물은 너무 끝없이 방대해서 방문하기에는 부담스럽다. 한시간 가량 지나면 지나온 공간이 잊혀 지기 시작하고 왜 건물 전체가 완만한 곡선과 최소의 디테일로 지어지지 않았는지 의문이 들기 시작한다.

# "IT IS AN ODD BUILDING TO VISIT, BECAUSE IT IS SO VAST AND RELENTLESS."

Rolex Learning Center sketch by OPARCH

# SLOT STUDIO

The roof of the Santa Caterina marketplace in Barcelona, Spain, designed by architects Enric Miralles and Benedetta Tagliabue, because the roof design recognizes the pedestrian level and also the urban presence and its effect on the immediate environs.

엔리크 미랄레스와 베네데타 태글리아부가 디자인한 스페인 바르셀로나의 산타 카테리나 시장의 지붕이다. 그 이유는 산타 카테리나 시장의 지붕 디자인은 보행자의 대지되며 인근에 대 으로 도시적 존재

© Radiuk
Santa Caterina marketplace

## Stefano Corbo Studio

The technological roof designed by Arata Isozaki at the Osaka Expo in 1970 has always fascinated me. When asked to design a Festival Plaza for the International Exhibition, Isozaki decided to reduce architecture to the definition of a gigantic and interactive roof, made of a metallic structure and of temporary capsules: these capsules were installed in order to display information and to exhibit the work of the most innovative designers.

1970년 오사카 엑스포에서 이소자키 아라타가 디자인한 기술적인 지붕에 매료되었다. 국제 전시를 위해 축제의 장 설계를 의뢰 받았을 때 이소자키는 거대한 규모의 움직이는 지붕 하나로 건축 구성을 축소시키기로 마음먹었다. 지붕은 철골구조와 임시 캡슐로 만들어졌다: 이 캡슐들은 가장 혁신적인 디자이너들의 작품과 프로필을 전시하기 위해 고안되었다.

© Process Architecture

Arata Isozaki, Festival Plaza, Osaka Expo, 1970, Japan

© Parpis Leelaniramol

### TOUCH Architect

The most memorable roof design we have seen, is at EWHA Womans University in Seoul, South Korea. It is the coffee shop building in front of the university museum, which is the thinnest roof shape we've ever seen. To create this kind of thin roof, it has to eliminate all the roof structure such as beams, in order to thinner the roof plate. This affected on the amount of the columns that should be more frequent than usual. A combination between the thin roof with plenty of small columns creates a new feeling and uniqueness, while still act as a functional roof which shade the space below.

우리가 본 지붕 중 가장 인상 깊은 지붕 디자인은 대한민국 서울의 이화여자대학교에 있다. 대학교 전시관 앞에 있는 커피숍 건물로 우리가 본 지붕 중 가장 얇은 지붕 디자인이었다. 이렇게 얇은 지붕을 만들기 위해서는 빔과 같은 지붕 구조물을 모두 제거해야 한다. 이 영향으로 평소보다 더 많은 기둥들이 세워져 있다. 얇지만 지붕 아래 그늘을 드리우는 지붕의 기능은 잃지 않으며, 수많은 작은 기둥들과 함께 만들어내는 조합은 새로운 분위기와 독특함을 자아낸다.

### TROPICAL SPACE

We love the beauty of sunlight in the caves. It is very mysterious and attractive to us.

우리는 동굴안으로 비치는 햇살의 아름다움을 사랑한다. 우리에게는 그 모습이 매우 신비롭고 매력적이다.

# Q5

Do you have a certain approach or your **own unique** touch when designing a roof? If so, what is the reason? If not, what kind of **signature design** would you make?

## b4 architects

Depending of the type of the project we take on. In general when the horizontal dimension is primary and the program is quite well-structured the roof element becomes a very important item. We like in these occasion to treat it as a whole that can resume all the underlying complexity trying to study a supporting system not located on edge exalting if it is possible an outdoor space covered by using some cantilever structure.

프로젝트에 따라 다르다. 기본적으로 수평적 형태로 꽤 구조화된 프로그램의 경우라면 지붕은 아주 중요한 아이템이 된다. 이런 경우 우리는 지붕을 그 아래 복잡성을 하나로 정리하는 전체로 간주하여, 지지구조를 지붕 가장자리에서 끝내지 않고 가능하면 캔틸레버 구조를 사용하여 외부 공간까지 커버하도록 디자인한다.

## BOARD

What I find interesting is when a roof is not merely perceived and used as a structural element and as a part of a building envelope that is covering the uppermost part of a building, but for other purposes that can be, for example, social or cultural. When we worked on the design for the extension of the "Vienna Museum" in Austria, we added several pavilions to the already existing ones in front of the museum as part of the extension. On the upper floor of one of these pavilions we located the so-called "Vienna-room", which is a flexible space that is publicly accessible and has a coffee bar on the upper floor and a roof terrace

건물의 가장 위 부분을 덮는 단순한 덮개 구조물로 인지되거나 사용되는 지붕이 아니라 사회 문화적인 면으로도 다양하게 활용되는 지붕을 좋아한다. 오스트리아의 "비엔나 박물관" 증축 작업에서 우리는 박물관 전면에 있는 기존 전시관에 여러 전시관들을 새롭게 추가하였다. 그 중 한 전시관 위로 이른바 "비엔나 실"이라고 불리는 방을 만들었다. 그 방에는 사람들이 자유롭게 드나드는 커피숍이 있고 주변 나무들 보다 더 높은 지붕 테라스에서는 광장과 도시가 내려다보인다.

that is located above the surrounding trees and overlooks the square and the city.

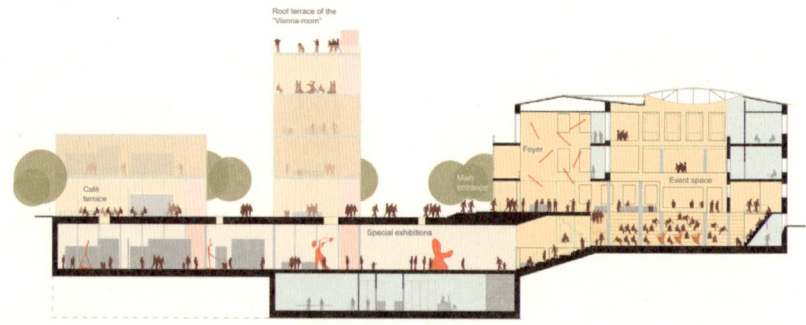

## ELA(Edu Lopez Architects)

As we have been saying in the previous questions, I like to create roofs that are different from the rest, that are not a simple horizontal element, but that they are part of the project and identify

앞선 질문에서 대답했듯이 우리는 단순히 수평적 요소가 아닌 프로젝트의 한 일부로 확실히 인지되는 차별되는 지붕을 만들기 원한다. 우리에게 지붕은 프로젝트의 가장 중요한 부

it. For me, the roof is one of the most important elements of the project because its geometry has to be identified inside and outside of the building. A couple of examples that we can give of this, is the cover of the art museum of Budapest and the Museum of the Bauhaus in Germany. In both projects it is seen that the roof is part of a compositional and project system of the project. The geometry that is seen in the exterior is in the marked interior being the form that acquires the same both inside and outside. In both projects it is observed that the cover is curved but in one of them it is concave and in the other convex.

분 중 하나로 지붕의 형태는 내부에서도 외부에서도 식별되어야 한다. 예를 들자면 부다페스트 미술관과 독일 바우하우스 박물관이 있다. 두 건축물의 지붕 모두 프로젝트 시스템의 한 구성요소로 인지되며, 외부에서 보여지는 형상은 내부에서도 동일한 모습이다. 보다시피 두 건축물에서 지붕은 똑같이 곡선처리 되었지만, 하나는 오목하고 다른 하나는 볼록하다.

Section 3

Section 4

Logically in the interior is inverted this geometry. The gallery of Budapest being a project of very large dimensions, the cover is replicating with different dimensions and angles of concavity to be able to accommodate a water system when it rains. However in the Bauhaus museum this deck is unique and convex to dislodge the water towards the perimeter of the building and create a greater height inside the large exhibition hall.

논리상 내부에서는 그 반대로 보일 것이다. 부다페스트 미술관은 매우 큰 규모의 건물로 여러 사이즈의 지붕으로 덮여 있으며 각도가 오목한 부분은 우천시 배수 시스템으로 활용된다. 반대로 바우하우스 박물관의 지붕은 독특 볼록 형태로 빗물을 건물 주변으로 흘러내린다. 지붕의 볼록한 부분의 내부는 다른 전시장보다 더 높게 뚫린 공간이 된다.

## Ezequiel Farca + Cristina Grappin

It depends on the location of the terrace, usually the context supplies the materials, we have used volcanic stone for several of our terrace projects in central Mexico, we like to use it for its functionality, it is perfect for open areas, can be easily obtained and the neutral chromatic gives us the possibility of generating different compositions with furniture, decoration objects and nature. There are projects when its context gives us special views, vegetation that we can integrate. It's all about context.

테라스의 위치에 따라 다르다. 대개는 주변환경에서 재료를 공수한다. 멕시코 중심부에 건물을 설계할 때 테라스에 화산석을 여러 번 사용했었다. 우리는 재료의 기능성을 살리는 걸 좋아한다. 화산석은 오픈된 공간에 안성맞춤이며. 구하기도 쉽다. 자연에서 유래된 색은 가구나 인테리어 소품 그리고 자연적인 요소등과 함께 다양하게 연출될 수 있다. 초목이 펼쳐진 환경은 지붕과 연계되어 프로젝트에 특별한 뷰를 선사하기도 한다. 지붕 디자인은 모두 지형적 조건과 연관된다.

## Katsuhiro Miyamoto & Associates

I believe that the roof is very important. It is a constant that hovers over the occupants lives and although additions and renovations are often done in buildings, the roof is usually a stable feature that lasts long.

지붕은 매우 중요하다고 생각한다. 공간 사용자의 생활 주변을 끊임없이 맴돌고 건물을 증축하거나 레노베이션을 자주 한다고 할지라도 지붕의 모습은 늘 오랫동안 변하지 않기 때문이다.

## Miha Volgemut architect

The roof is the most dominant feature on the house and it gives the character to it, that's why I am not such a big follower of flat roofs. When you want to be creative; the roof design is probably the vastest playground for architect.

지붕은 주거건물에서 가장 지배적인 특색을 가지며, 건물에 개성을 불어넣는다. 그렇기 때문에 나는 그리 평평한 지붕의 옹호자가 아니다. 창의적이고 싶은가; 아마도 지붕 디자인이 건축가에게 가장 방대한 놀이터가 될 거다.

## Moussafir Architectes

If so, what is the reason? If not, what kind of signature design would you make? Not really, it all depends on the project.

특별히 없다. 프로젝트에 따라 다르다.

## NL Architects

We like to consider the roof as the 'fifth' façade, so it is worthwhile to threat it as something precious, with great potential. In dense environments a (flat) roof forms a valuable surface. It is a good idea to look for programmatic use beyond providing

우리는 지붕을 "제5의" 입면으로 보고 싶다. 그래서 지붕에는 정성을 쏟을 만한 가치가 있으며 엄청난 잠재력이 있다. 복잡한 환경에서 (평평한) 지붕은 그만큼 가치가 높다. 이럴 경우 비바람을 피해주는 단

shelter. It is not so much about signature, but about functionality and fun. I for instance love the ubiquitous typology of usable flat roofs on Jeju Island that are accessible via external stairs that wrap around the houses in a sculptural way.

순한 공간을 넘어서 실용적인 활용도를 모색하는 것이 좋다. 상징성만이 아니라 기능과 재미를 말하는 것이다. 예를 들어 나는 제주도에서 흔히 볼 수 있는 집 주변을 두르는 외부 계단을 통해 마치 조각 하듯 접근하여 활용할 수 있는 평평한 지붕을 좋아한다.

## object-e architecture

A main principle of our design methodology and ethos is that each project is unique and has to answer to the 'questions' that it deals with each time. Therefore we try to avoid any idea of 'style' that would be imposed on a design from the 'outside'. In that context, our design of roofs follows also a unique approach each time (excluding maybe the houses on the Cycladic islands mentioned in the previous question). The following two projects can illustrate that.

Our proposal for the passage terminal at the port of Souda near Chania in Crete, has as its starting point the typology of the 'Arsenalia'. Arsenalia, a 15th century structure used to repair ships, can be found in the city of Chania and define its architectural character. They consist of a long volume with a cylindrical roof. Our design acknowledge that heritage, and given that the new building has an equally important relation to the sea as the Arsenalia had, derives out of an abstraction of their basic scheme: the repetition of semicircles is transformed into a modern structure evolved through the abstraction of the idea. The building is made out of exposed concrete, while copper and glass complement its material

우리 디자인 방법과 철학의 주요 원칙은 각각의 프로젝트마다 고유성을 가져야 하며 시시때때로 마주치는 '의문점'을 해결 할 수 있어야 한다는 것이다. 그래서 우리는 '외부'에 비춰지는 그 어떤 '스타일적'인 생각을 버리려고 노력한다. 이런 취지로 설계하는 지붕은 모두 그만의 방식이 있다. 다음의 두 프로젝트가 이를 설명해준다.

크레타 섬 하니아 현 근처에 있는 수다 항구 터미널을 위한 설계안은 'Arsenalia(아르세날리아)' 유형학에서 출발하였다. 'Arsenalia'는 15세기 배를 수리할 때 사용했던 구조체로 하니아 도시에서 볼 수 있으며 이 지역의 건축적 특성으로 자리잡았다. 우리는 Arsenalia의 기본 방식이 함축된 디자인적 지식유산을 새 건축물에 적용하였으며, Arsenalia가 그랬듯이 새 건축물도 바다와 중요한 관계를 맺고 있다. 반복되는 반원 체 원리가 함축되어 현대적 구조로 진화하였다. 노출 콘크리트로 지어진 건축물에 구리와 유리소재를 덧붙여 재료적 개성을 부여했다. 건물의 지붕은 전통 지붕 시스템을 현대적으로 해석한 결과물로 전체적으로 건물의 체계를 형성한다.

vocabulary. Therefore the roof of the building is a modern interpretation of a traditional roofing system and becomes the defining scheme for the whole project.

Souda Port Terminal

On a different direction, ballRoom is an 'architectural folly'. Is a project designed for the MC Redux Exhibition, organized at an abandoned building at the waterfront of Thessaloniki, Greece; a concrete structure echoing the principles of modern architecture. The curators asked the public to propose possible uses for the building through a facebook page, and then invited several architects to choose one and design a quick proposal. ballRoom is our take on the theme of the exhibition.

In the above context, ballRoom attempts to bring a sense of the unexpected to the waterfront of the city. It consists on an interweaving of linear arcs that emerge through the self-organization of linear elements. Their form is derived from a digital simulation of the 'hanging chains model'. In essence our proposal is a new roof that takes the place of the existing one and totally alters the character of the existing building. In the interior of it the inverted arcs penetrate the roof of the abandoned shell, which itself represents a long gone modern ethos, and form a proper ball-room. In the outside of the building they create some kind of 'grottos'; rather dark and protected spaces that try to introduce a sense of privacy and

다른 한편에는 '건축적 장식'인 ballRoom 사례가 있다. ballRoom은 MC Redux 전시회를 위해 설계된 프로젝트로 전시회는 그리스 테살로니키의 해안가에 버려진 건물에서 개최되었다: 근대 건축의 원칙을 그대로 따라한 콘크리트 건물이다. 큐레이터는 대중에게 페이스북으로 건물의 활용 방안에 대한 제안을 요청하였고, 여러 건축가들을 초청해 그 중 하나씩 선택하여 즉각적으로 디자인하도록 하였다. ballRoom은 전시회 주제에 맞게 우리가 택한 안이다. 우리는 ballRoom을 통해 도시 해안가에 예상밖의 감각을 심어 주려는 시도를 하였다. 선적인 요소들이 자기조직화를 통해 활 형태 짜임으로 구성되었다. 형태는 'hanging chains model' 디지털 시뮬레이션을 통해 만들어졌다. 프로젝트의 중심에는 원래 지붕이 있던 자리에 그대로 존재하지만, 건물의 기존 성향을 완전히 바꾸어 버리는 지붕이 있다. 지붕의 내부는 거꾸로 휘어진 활형태가 오래전에 잊혀진 근대 철학을 대표하는 낡은 지붕 구조를 뚫고 지나가 무도회장(ball-room)에 적합한 환경을 조성한다. 건물 외부에서는 도시 중심으로부터 격리되어 약간은 어둡고 사생활이 보호될 것 같은 안전한 공간의 '작은 동

isolation in the public space of the city. Both in terms of form, as much as of function, ballRoom is in contrast with the main characteristics of the public spaces and buildings of Thessaloniki: In contrast with open green spaces where everything is visible and transparent; In contrast with rectilinear shapes and volumes and clean lines; In contrast with an understanding of architecture as something clean and sterile. ballRoom is therefore a roof that has no relation to traditional types of roof (where the modern flat concrete roof is understood as an already 'traditional' type) and becomes an elaborate and exuberant web of lines that merge ornament with structural elements.

굴'과 같은 효과를 낸다. 기능과 그에 못지않게 중요한 형태 두 측면에서 본다면 ballRoom은 테살로니키에 있는 건축물과 공공 장소의 주요 특성과는 대조된다. 모든 것이 훤히 보이는 오픈 된 녹지 공간과 대조된다. 살균한 듯 깨끗한 것으로 인지되는 건축물들과는 대조된다. ballRoom의 지붕은 전통 지붕 타입(근대적 평평한 콘크리트 지붕이 이미 '전통적' 타입이라고 간주된다)과는 아무런 연관성이 없는 지붕이며, 구조적 요소와 장식적 요소가 통합된 정교하면서도 활기 넘치는 거미줄 형태의 지붕이다.

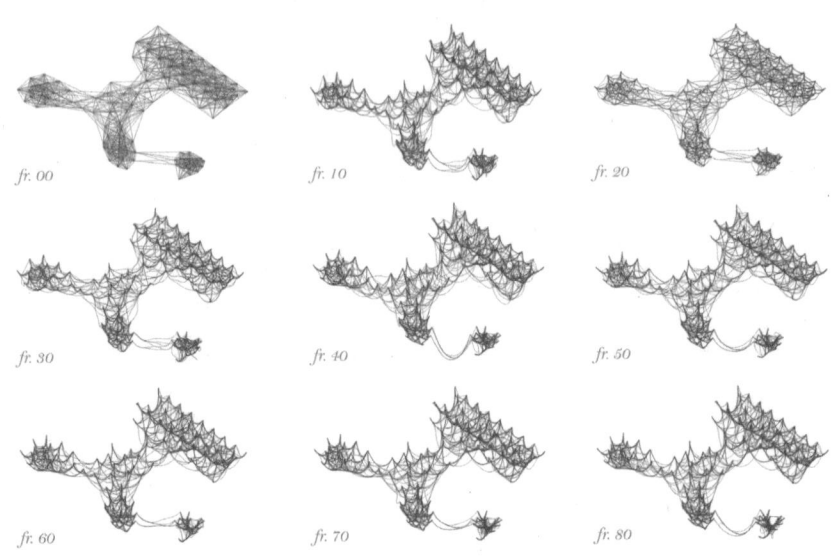

**ballRoom – form generation of the roof / steps**

ballroom

## OPARCH

Ultimately, the thing I care most about in architecture is space. Building elements (walls, roofs, floors, etc.) are simply raw materials to create spatial effects. Currently, I am very interesting in dynamic, swirling spaces. So the short answer is 'yes', insofar as any roof must be integrated with an overall composition.

근본적으로, 내가 건축에서 가장 신경 쓰는 것은 공간이다. 건물의 요소들(벽, 지붕, 바닥 등)은 단지 공간적 효과를 내기 위한 소재일 뿐이다. 현재 나는 역동적이고 소용돌이치는 공간에 관심을 가지고 있다. 그래서 모든 지붕이 전체적인 구성과 통합되어야 한다면 이 질문의 대답은 간단히 '그렇다'이다.

Dune

## SLOT STUDIO

The design approach for roofing systems Slot employs takes into account three fundamental elements: the roof's correspondence to the totality of the architectural project, its accessibility in

저붕 시스템을 위한 우리 디자인적 접근에는 3가지 기본적인 요소가 있다. 아래의 내용이 바로 그것이다. "건물 전체의 통일성에 부합해야 한다. 디자인적으로 이해가

terms of design and the use of natural light to enhance and supplement interior lighting.

쉬워야 한다. 내부 조명 효과를 높이거나 지원하기 위해 자연 조명을 사용한다."

AZTEC STADIUM

HAKKA CULTURAL CENTER

House of John Paul

## Stefano Corbo Studio

I always try to consider the roof not as a horizontal elevation, but as a layer, or a floor, which is completely integrated into the life of the building I'm designing.
For this reason, my aim is often to include the roof in the main circulation system of my projects; or, alternatively, to assign to the roof a specific function (public space, thermodynamic device, etc.).

나는 지붕을 단지 수평고도가 아닌 완전히 건물의 심장과 융합되는 하나의 레이어 또는 층으로 간주하려고 노력한다. 그렇기 때문에 나는 지붕을 건물 주요 순환(통로) 시스템 안으로 많이 포함시키는 것이 목표다; 그것이 여의치 않으면, 지붕에 특정 기능을 부여한다. (공공장소 혹은 열역학의 장치 등…)

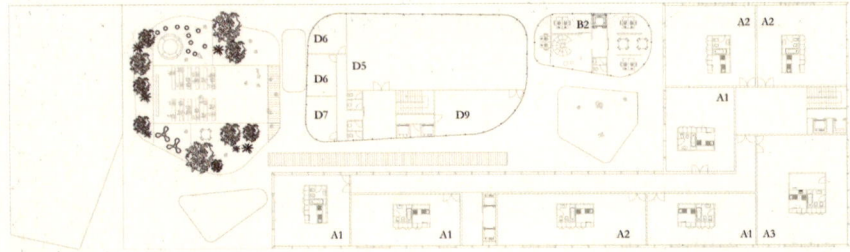

Arts Center Roofplan

## TOUCH Architect

When designing an architecture, it has to be started with function and context. So, there is no pattern of our design style, in other words, we are more concern about the most suitable roof or other elements which befit in each architectural design.

설계는 기능과 문맥에서 출발해야 한다. 그렇기 때문에 우리의 설계 스타일에는 일정한 패턴이 없다. 즉, 우리는 각 프로젝트를 위한 가장 적합한 지붕 등 가장 적합한 디자인 요소를 찾는 것을 더 중시한다.

햇빛은 대부분 틈새가 있는 벽 돌 벽의 일부인 지붕을 통해 얻는데, 다양한 빛의 효과가 나타난다.

"WE USUALLY GET THE SUNLIGHT FROM THE ROOF, GOING WITH BRICK WALL WITH HOLES CREATE DIFFERENT LIGHTING EFFECT."

TROPICAL SPACE

# Q6

Are there **any myths, stories or superstitions** about roofs in your country?

# GREECE
## object-e architecture

Interestingly, roofs are considered in the oral tradition of Greece as ideal places for famous architects to end the rivalry with their students that thread to become better than them. The first occurrence of the theme can be found the story of Daedalus, who is in fact the archetypical figure of the architect in Greek mythology. Daedalus, while still in Athens, took as an apprentice his nephew, Talus. Talus, apparently proved to be quite talented: Among other things he was supposed to invent the saw, after observing the spine of a fish (or the teeth of an animal according to a different version). Daedalus, afraid that his nephew will overshadow him, led him to the roof of the temple of Athena in the Acropolis, supposedly in order to show him the view, but instead pushed him over the edge and killed him. The murder of Talus was the reason Daedalus was forced to leave Athens and find refuge in Crete, where he made some of his most famous works, including the labyrinth.

A second appearance of the same theme can be found in the story of the architect of the restoration of the Panagia Ekatontapyliani church, on the island

재미있게도 그리스 구전에 의하면 지붕은 유명한 건축가들이 자신을 뛰어넘을 기미가 보이는 제자들을 상대로 경쟁을 끝내기에 이상적인 장소라고 한다. 이와 연관된 첫 사건은 실제 그리스 신화에서 전형적인 건축가의 모습이었던 다이달로스의 이야기에서 찾을 수 있다. 다이달로스가 아테네에 머무를 동안 그의 조카 탈로스를 제자로 삼았다. 탈로스는 상당히 재능이 있는 듯했다. 여러 업적 중 탈로스는 생선의 척추 뼈를 관찰한 후 (또 다른 버전에서는 동물의 이빨을 관찰한 후) 톱을 발명하기도 했다. 조카의 그늘에 가려질 것을 두려워한 다이달로스는 경치를 보여준다는 핑계로 조카를 아크로폴리스의 아테나 성전 지붕 위로 불러냈다. 하지만, 경치를 보여주는 대신 지붕 끝에서 조카를 밀어 죽였다. 탈로스를 살인한 댓가로 다이달로스는 아테네에서 쫓겨나 크레타 섬으로 피신하였다. 거기서 그는 미노스 왕의 미궁(라비린토스) 등 그의 가장 유명한 작품들을 남겼다. 동일한 내용의 사건으로 비잔틴 제국의 유스티니아누스 1세 때 파로스 섬의 Panagia Ekatontapyliani 교회 복구작업을 한 건축가의 이야기가 있

111

of Paros, during the Justinian times of the Byzantine Empire. The church was allegedly founded by Saint Helen, the mother of the first Byzantine emperor, Constantine. It was restored and greatly expanded during the reign of the emperor Justinian I. According to the tradition, the architect of the new temple was a student of the architect of Hagia Sophia, the cathedral of Constantinople, center of the empire. This time around, it was the student who invited the teacher to the roof, proud of the new dome that he constructed. The teacher, jealous of the achievement of his student followed Daedalus' footsteps and pushed him over the edge to his death. Of course, the dome of Hagia Sophia is by far the most impressive achievement of Byzantine architecture while the one of Panagia Ekatontapyliani is relatively small. Nevertheless, the story remains as another example of this 'ancient' method of countering competition from fellow architects.

다. 전해진 바에 의하면 교회는 첫 비잔티움 황제인 콘스탄티누스의 어머니 세인트 헬레나에 의해 세워졌다고 한다. 교회는 유스티니아누스 1세의 통치기간동안 복구되었고 크게 확장되었다. 구전에 따르면 교회를 복구한 건축가는 제국의 중심인 콘스탄티노플에 위치한 하기아 소피 대성당을 지은 건축가의 제자였다고 한다. 이번에는 반대로 제자가 자신이 설계한 새 돔을 자랑하기 위해 스승을 지붕으로 불렀다. 스승은 제자의 성취를 시기하였고, 다이달로스의 전례를 밟아 제자를 지붕 끝으로 밀어 죽게 하였다. 당연히 더 멀리 보이는 하기아 소피아 대성상의 돔은 비잔틴 건축에서 가장 위대한 업적이었다. 이에 비하면 Panagia Ekatontapyliani 교회의 돔은 빈약하기 그지 없었다. 그럼에도 불구하고, 이 이야기는 동료 건축가와의 경쟁에 대응하는 '고대적' 방법의 또 하나의 사례이다.

## ITALY
### b4 architects

In our language, Italian, there are many way of saying or expression in which the

우리 이탈리아어에는 지붕을 뜻하는 단어나 지붕을 표현 방법이 아주 많다. 지붕에 대한

element of the roof is present. The most known superstition about roof is when you open an umbrella inside: this is a great sign of misfortune perhaps because it evokes a broken roof or worse the lack of the roof and once this was a sign of misery.
On the contrary when a swallow chooses to make its nest under your roof is a sign of good luck.

가장 잘 알려진 미신으로 지붕 아래에서 우산을 펼치면 지붕이 붕괴 되었거나 지붕이 모자란다는 뜻으로 고통을 동반한 엄청난 불행이 따른다고 믿는다.
반대로 참새가 지붕 밑에 둥지를 틀면 행운이 온다고 믿는다.

## ITALY
### Stefano Corbo Studio

A very old superstition about roofs deals with the presence of owls. Owls, especially in the Greek and Roman tradition, were associated with witches, and more in

지붕에 대해 아주 오래전부터 전해오는 미신은 부엉이의 존재와 관련 있다. 특히 그리스와 로마 전통에서 부엉이는 마녀나 더 일반적으로는 악마와

general, with Evil. So if an owl perches on the roof your house it will bring you bad luck, or even death.

연관된다. 그래서 당신의 집 지붕위에 부엉이가 앉아 있다면, 당신에게 불운이 오거나 죽음을 맞이 할 수도 있다.

## JAPAN
### Katsuhiro Miyamoto & Associates

In Japanese religious beliefs, it is often that a roof is capped over objects of significance as a gesture of gratitude or prayer to God.

일본 종교에서 지붕은 주로 신에게 취하는 감사나 기도와 같은 중요한 행위를 보호해주는 요소로 간주된다.

© Katsuhiro Miyamoto & Associates
Roof capped onto lightning struck tree

# MEXICO
## Ezequiel Farca + Cristina Grappin

Maybe, to hear footsteps on the roof. It's about a person who feels threatened because it has been activated a state of guilt in consciousness, the origin of this myth is not known, but it makes the physical analogy with a roof because of the feeling of threat for something supreme, something that It is on you, and of which you can not have concrete knowledge because it is not possible to see.

아마도 지붕위에서 발자국 소리가 들린다는 것은 누군가가 양심의 가책을 느껴 불안한 심정에 있다는 의미일 것이다. 어떻게 이 이야기가 유래되었는지는 모르지만, 자신보다 우월한 무언가로부터 위협을 느낀다는 것이 내 머리 위에 있고 보이지가 않아 정확하게 알 수 없는 지붕과 물리적으로 연관된다.

# MEXICO
## SLOT STUDIO

The ancient building tradition used in Maya houses reflects many beliefs held

마야인의 주거에 사용되었던 고대 전통 건축은 많은 일상적

in daily life and in Maya theology. For example, the round roof represents the celestial vault and the wooden roof posts always number fifty-two, which is an allusion to the life cycle of the Maya.

종교의식과 마야 신학을 반영한다. 예를 들어 돔 형태의 지붕은 천상의 아치천장의 결과물이며 목재 지붕 뼈대는 마야 사람들의 생활주기를 암시하는 숫자를 반영하여 언제나 52 개이다.

© Gildardo Sanchez

## NETHERLANDS
### NL Architects

Sinterklaas is a Spanish holy man of about 500 years old. On his birthday, the 5th of December, De Sint delivers presents by traveling over the rooftops with his horse throwing the gifts through the chimneys.

성 니콜라스(Sinterklaas)는 약 500살 정도 되는 스페인의 성자이다. 성 니콜라스는 그의 생일인 12월 5일에 말을 타고 지붕위를 넘어다니며 굴뚝으로 선물을 전해준다.

## NETHERLANDS BOARD

Since 2015 Rotterdam every year organises the so-called "Rotterdamse Dakendagen", which can be translated as "Rotterdam Rooftop Days". It is a three-day festival that celebrates the roofs of Rotterdam's buildings by making many of them publicly accessible to stimulate their use and development. The idea to create an annual event on the roofs of Rotterdam developed out of a mini-roof festival called "Poems from a Rooftop" that took place in 2013 as part of a music festival, which included the use of rooftops for bars and organized excursions to roofs. Today it is a festival that provides public access to more than forty roofs for exciting discoveries, informal drinks, intimate concerts, silent discos, film nights, sports activities, and children's programmes.

2015년부터 로테르담은 매년 소위 "로테르담 옥상의 날"이라는 뜻을 가진 "Rotterdamse Dakendagen"라고 불리는 행사를 개최한다. 3일동안 로테르담의 건물 지붕들을 기념하는 페스티벌로 지붕의 활용과 개발을 도모하기 위해 대중들에게 많은 지붕들을 개방한다. 이 로테르담의 지붕을 위한 연간 행사는 2013년 뮤직페스티벌의 일부로 지붕을 바(술집) 등 다른 용도로 활용하는 옥상의 와도 행사였던 "옥상에서 전해오는 시"라는 미니-루프 페스티벌에서 비롯되었다. 요즘날에는 40여개 이상의 지붕을 대중에게 개방되어 체험장, 캐주얼한 술집, 작은 콘서트, 싸일런트 디스코, 영화의 밤, 스포츠 게임 그리고 어린이 프로그램 등의 행사가 열린다.

© G.Lanting

## SLOVENIA
### Miha Volgemut architect

I don't think there is a particular myth related to the roof element in my country, but for example it is a custom in my country to put the small spruce tree on the top of the roofing when built. People believed that evergreen tree has a special life power; the spruce than protects the home and brings vitalism, health and happiness into the house.

우리나라에는 지붕디자인에 관해 특별히 전해 내려오는 이야기는 없는 것 같다. 하지만, 우리나라에서는 집을 지을 때 지붕 위에 작은 가문비나무를 심는 관습이 있다. 사람들은 상록수가 특별한 삶의 힘을 가지고 있다고 믿어왔다; 그런 의미로 가문비나무는 집을 보호해주고 가정에 활기와 건강 그리고 행복을 가져다 준다고 믿는다.

## USA
### OPARCH

I always find it odd that many Americans associate traditional eaves with 'friendliness'. The equation seems to be more vernacular = more friendly. I have read various academic essays claiming this tendency has to do with a reaction against modernism triggered by a fear of European sophistication. Perhaps, but whatever the cause, this fear of abstraction is real.

나는 많은 미국인들이 전통 처마를 '친근감'과 연관시키는 것이 이상하다. 등식은 이런 것 같다. 더 토착적인 것 = 더 친밀한 것. 나는 이러한 경향은 유럽의 세련됨에 대한 두려움에서 촉발된 모더니즘에 대한 반발과 연관성이 있다고 주장하는 여러 학술논문을 읽은 적이 있다. 무엇이 원인이던 간에 아마도 이러한 추상적인 것에 대한 두려움은 실제일지도 모른다.

A Rain-falling Temple

## THAILAND
## TOUCH Architect

As we know, there are no myths, stories, or superstitions about roof in Thailand. We only concern about climate affect which is hard rain all the year.

태국에는 지붕에 대한 신화나 이야기 또는 미신이 없다. 우리는 한 해 동안 폭우가 얼마나 내릴지에 대한 기후 영향에만 신경을 쓴다.

## VIETNAM
## TROPICAL SPACE

We have old quote in Vietnamese, it could be translated as: "a child without father is like a house without the roof."

베트남에는 예부터 내려오는 이런 문구가 있다. "아버지가 없는 어린이는 지붕이 없는 집과 같다."

b4 architects

BOARD

ELA(Edu Lopez Architects)

Ezequiel Farca + Cristina Grappin

Katsuhiro Miyamoto & Associates

LGSMA_

Miha Volgemut architect

Moussafir Architectes

NL Architects

object-e architecture

OPARCH

SLOT STUDIO

Stefano Corbo Studio

TOUCH Architect

TROPICAL SPACE

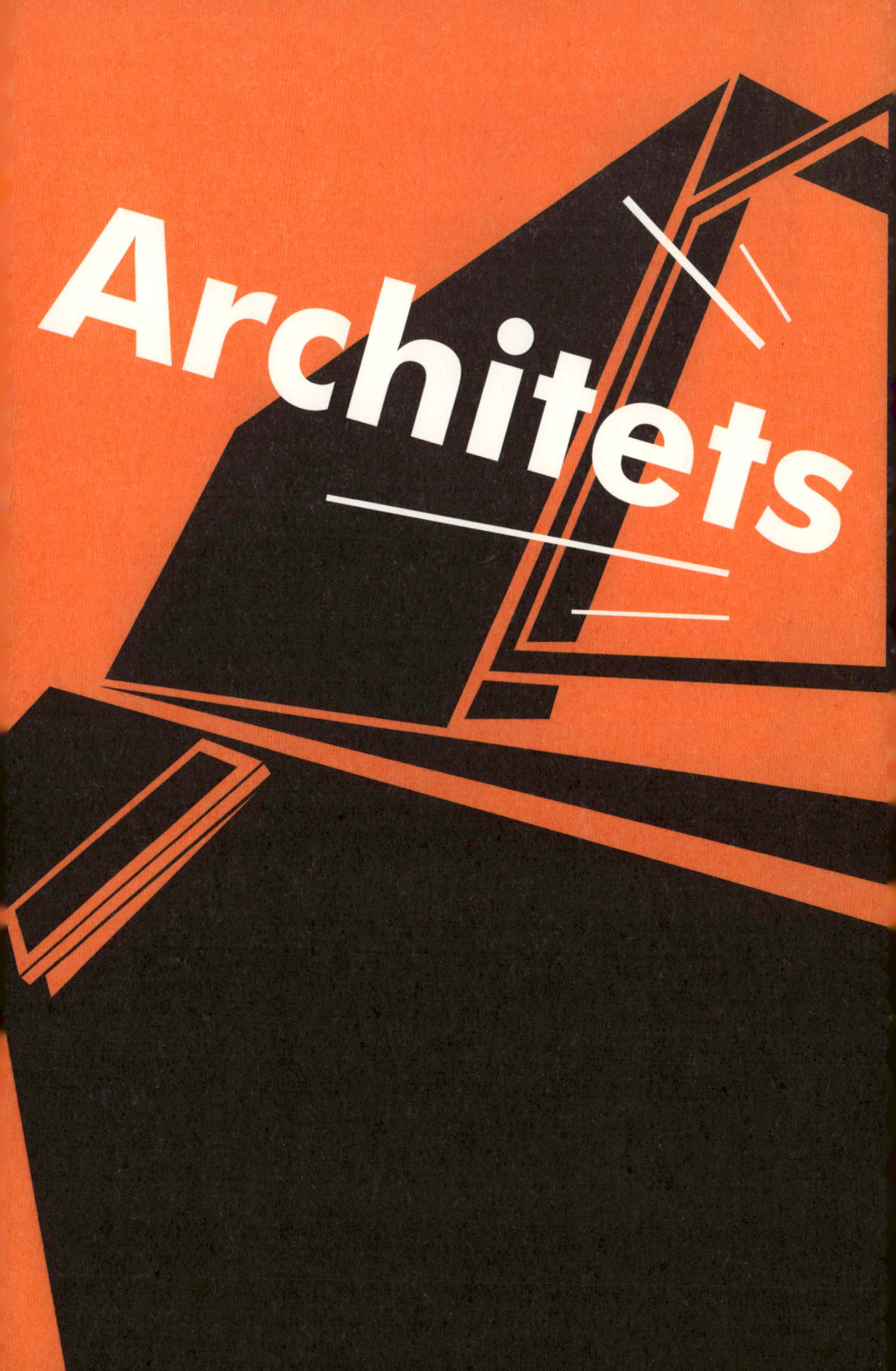

### b4 architects

"We are interested in producing works that contribute to the debate of the complexity of modern life."

### BOARD

"BOARD won several prizes recently in prestigious international architecture and urban design competitions."

### ELA-Edu Lopez Architects

"ELA endeavors to obtain a deeper beginning in the experience of time, space, light and materials."

**Ezequiel Farca + Cristina Grappin**

*"the perfect synergy between design and functionality"*

**OFFICE INFO**
4-29, Yumoto-cho, Takarazuka, Hyogo, 665-0003, JAPAN
kmaa@kmaa.jp

**Katsuhiro Miyamoto & Associates**

**LGSMA_**

"Architecture is seen as a test field of ideas and very different research tools, which ultimately aims at the Project in every kind of form."

**Miha Volgemut architect**

"Since May 2009 he is registered architect in the Chamber of Architecture and Planning Slovenia."

© Matevž Maček

**Moussafir Architectes**

"For us, architecture is less a question of image than of 'sense', in both meanings of the word."

**NL Architects**

"We understand architecture as the speculative process of investigating, revealing and reconfiguring the wonderful complexities of the world we live in."

"Object-e is based on several collaborations with people coming from different backgrounds, with different design intentions and agendas."

**object-e architecture**

## OPARCH

"In all the work, there is an emphasis on communicating architectural meaning by creating powerful emotional and perceptual resonances."

## Stefano Corbo Studio

"Stefano founded his own office, a multidisciplinary network practicing architecture and design, preoccupied with the intellectual, economical and cultural context."

## SLOT STUDIO

"As architects, we push design to its ultimate material consequences and aim for cultural connectivity."

**TOUCH Architect**

" Mr. Setthakarn Yangderm as an architect and leader of our firm and Ms. Parpis Leelaniramol as an architect, which will corporate together in design, construction, and management."

" Architecture with simple shapes, focus on ventilation solutionand natural lighting which is suitable with the tropical climate."

**TROPICAL SPACE**

# CASE
## STUDY

© Tomaz Gregoric, Jan Celeda

**2**

1. **BAROQUE COURT APARTMENTS**
   OFIS ARCHITECTS

2. **BEACH HOUSE**
   Lima Urban Lab

3. **BB HOME**
   H&P Architects

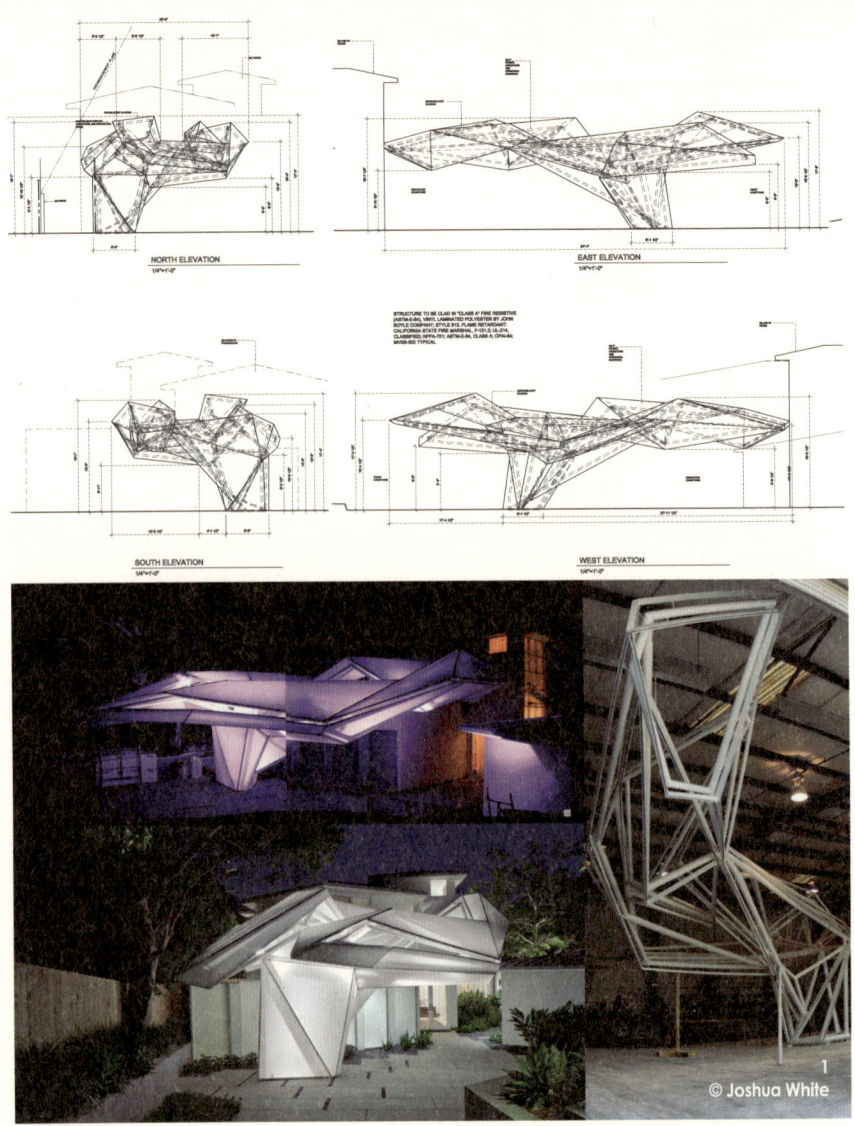

1. **Frank / Kim Residence**
   B+U

2. **Gyeongbokgung-Gangnyeongjeon eaves**

134  ROOF, protected Open Space

1. **PEAK SERIES**
   visiondivision

2. **RED+ HOUSING MANIFESTO**
   OBRA Architects

3. **Samri Residence_Hyojae**
   Koo Seungmin

1

2

136  ROOF, protected Open Space

1 | **Abrisham Pedestrian Bridge** & **Flower Market**
DIBA Tensile Architecture

© Mohammad Hassan Ettefagh

140 ROOF, protected Open Space

1. **Gyeongbokgung Jabsang**

2. **Dongdaemun Plaza Kiosk**
   NL Architects

3. **La Dynamo de Banlieues Bleues**
   PERIPHERIQUES

1. **Markthäuser 11-13**
   FUKSAS

2. **R4**
   Florian Busch Architects

3. **Tajiki Residential Tower Canopy**
   DIBA Tensile Architecture

# 1. Gyeongbokgung-Geunjeongjeon Area

1. **Ozawa dental clinic**
   DAI NAGASAKA / MÉGA

2. **ALTAMIRANO WALKWAY**
   EMILIO MARIN architect

3. Changdeokgung–Buyongjeong

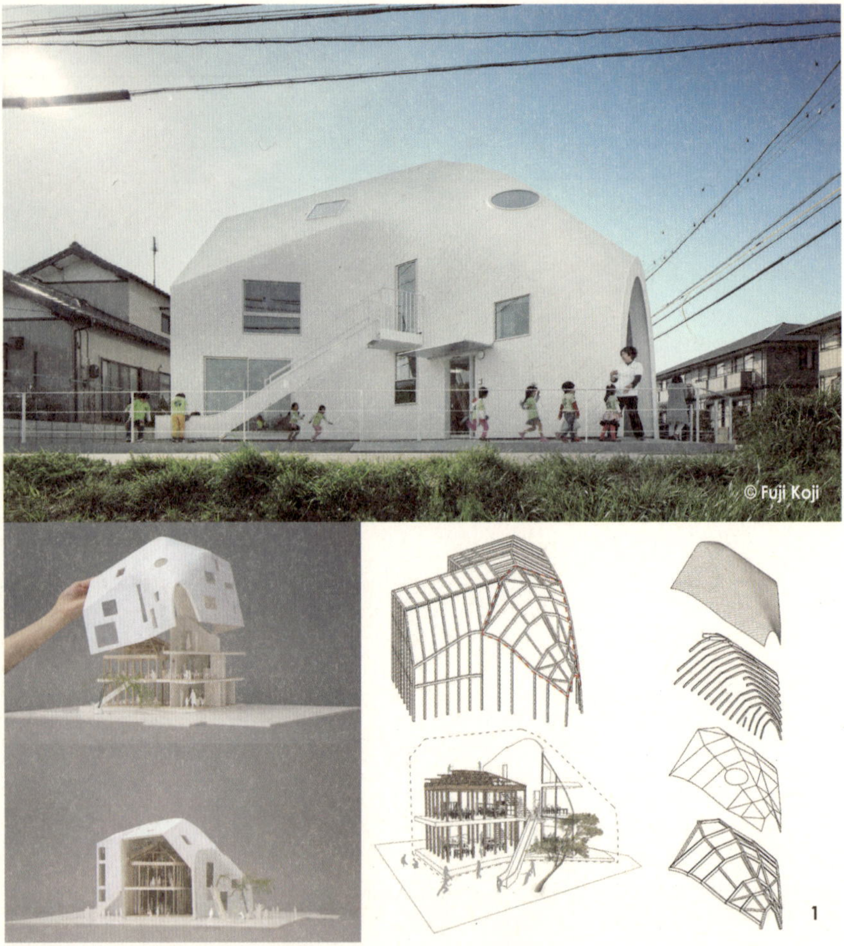


1

2

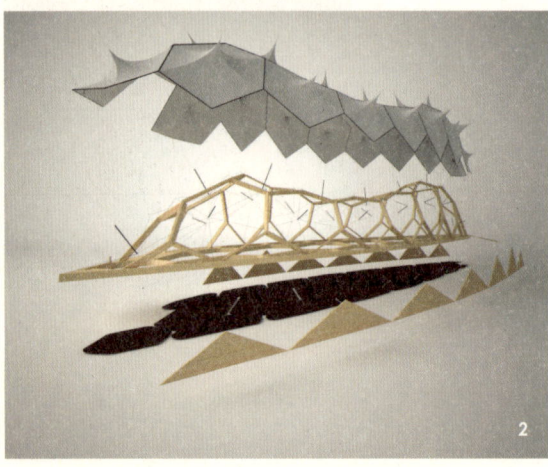

1. **Clover House**
   MAD Architects

2. **CUTTY SARK PAVILION**
   BAKOKO

3. **Basket Bar**
   NL Architects

1. Magoksa
2. Buseoksa
3. Haeinsa Janggyeong Panjeon

152 ROOF, protected Open Space

1. **Dalseong Citizen's Gymnasium**
   BXBstudio Bogusław Barnas

2. **Deoksugung-Hwangudan**

3. **FLYING PAVILION**
   3GATTI

4. **Gwanghanru**

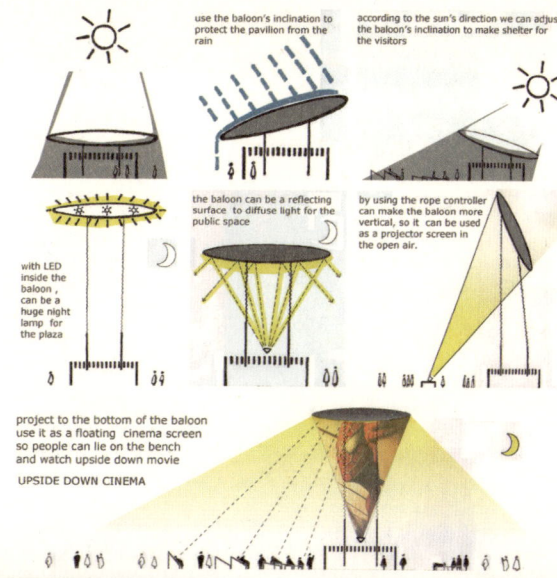

1. **IDYLL TOWER**
   Ryszard Rychlicki

2. **Library Technical University Delft**
   mecanoo

154  ROOF, protected Open Space

© Harry Cock

1. Changdeokgung

1. **Admirant Entrance Building**
FUKSAS